18th International Scientific and Technical Conference Rapid Solidification Materials and Coatings (RSMC 2021)

AA007326

Journal of Physics: Conference Series Volume 2144

Moscow, Russia
19-20 October 2021

ISBN: 978-1-7138-5115-8
ISSN: 1742-6588

Printed from e-media with permission by:

Curran Associates, Inc.
57 Morehouse Lane
Red Hook, NY 12571

Some format issues inherent in the e-media version may also appear in this print version.

This work is licensed under a Creative Commons Attribution 3.0 International Licence.
Licence details: http://creativecommons.org/licenses/by/3.0/.

No changes have been made to the content of these proceedings. There may be changes to pagination and minor adjustments for aesthetics.

Printed with permission by Curran Associates, Inc. (2022)

For permission requests, please contact the Institute of Physics at the address below.

Institute of Physics
Dirac House, Temple Back
Bristol BS1 6BE UK

Phone: 44 1 17 929 7481
Fax: 44 1 17 920 0979

techtracking@iop.org

Additional copies of this publication are available from:

Curran Associates, Inc.
57 Morehouse Lane
Red Hook, NY 12571 USA
Phone: 845-758-0400
Fax: 845-758-2633
Email: curran@proceedings.com
Web: www.proceedings.com

TABLE OF CONTENTS

LASER, ELECTRON-BEAM AND ION-PLASMA METHODS OF SURFACE MODIFICATION

Overview of the 18th International Scientific and Technical Conference Rapid Solidification Materials and Coatings

A A Lozovan[1] and D K Kostrin[2]

[1] Department of Materials and Materials Technology, Moscow Aviation Institute (National Research University) "MAI", 121552, Orshanskaya, 3, Moscow, Russia
[2] Department of Electronic Instruments and Devices, Saint Petersburg Electrotechnical University "LETI", 197376, Professora Popova, 5, Saint Petersburg, Russia

E-mail: alexandr.lozovan@yandex.ru

Abstract. In this preface the main results of the 18th International Scientific and Technical Conference Rapid Solidification Materials and Coatings that was held on 19–20 October 2021 in Moscow, Russia are discussed.

From 19 to 20 October 2021, the 18th International Scientific and Technical Conference Rapid Solidification Materials and Coatings (RSMC–2021) was held at the Moscow Aviation Institute (National Research University).

Despite the difficulties associated with the coronavirus pandemic, more than 100 specialists from leading universities, the Russian Academy of Sciences and Russian enterprises, as well as scientists from Ukraine and Belarus, took part in the Conference. For the successful holding of the Conference in a pandemic, all the necessary measures were taken to ensure the implementation of sanitary rules and standards for the protection of Conference participants.

At the Conference the results, which were obtained in the last year after the previous Conference, of the theoretical and experimental studies on the physics of condensed matter, low-temperature plasma physics, electronic and ion-plasma surface modification, physical methods of film deposition and other studies obtained over the last year after the previous Conference.

The main objects of these studies were superficiality and superficial phenomena. Therefore, to the development and improvement of methods of surface modification and coatings deposition, as well as methods of their research and development of modern surface treatment equipment was paid the greatest attention at the Conference.

During the Conference more than 70 reports were presented, which were heard in the following, traditional Conference sections (figure 1):

- section 1 – Rapid solidification materials;
- section 2 – Methods of coating application and research;
- section 3 – Laser, electron-beam and ion-plasma methods of surface modification;
- section 4 – Composite materials and permanent joints;
- section 5 – Works of young scientists.

The Conference was opened by Alexander Vladimirovich Bespalov, Co-Chairman of the Director of the Department of Materials Science and Materials Technology of the Moscow Aviation Institute (National Research University), who emphasized the importance of developing research in the field of

Content from this work may be used under the terms of the Creative Commons Attribution 3.0 licence. Any further distribution of this work must maintain attribution to the author(s) and the title of the work, journal citation and DOI.
Published under licence by IOP Publishing Ltd

physical and physicochemical methods of obtaining and processing new materials and coatings for the aerospace industry. He also noted the achievements of the Institute's scientists in the field of laser coating and fast solidification materials.

Figure 1. Working sessions of the Conference.

Plenary reports of the Conference were devoted to theoretical and practical aspects of microarc oxidation of Al and Mg alloys, mechanisms of formation and properties of nanocomposite coatings based on oxides of zirconium, hafnium and aluminum in plasma processes.

Of interest is a new approach to studying the features of the propagation of quasi-monochromatic optical fluxes through a thin flat transparent layer of a multilayer coating. It is shown that these fluxes can be transported by a layer in the course of its multiple sequential total internal reflection or by a waveguide-resonant propagation method. A combined surface treatment of metals by electrolytic-plasma diffusion saturation and polishing was also presented.

A number of reports presented the continuation of research on the use of inverted magnetrons for the deposition of composite multilayer coatings on substrates of complex shapes. Further development of the method for visualizing the deformation of materials arising under the action of laser pulses using mechanoluminescent coatings is proposed. A number of reports were devoted to the continuation of research in the field of microarc oxidation.

New approaches to the description of electronic and proton processes of multiple scattering in solids are presented.

Considerable attention was paid to reports on the formation of nitride protective coatings by vacuum-arc and magnetron methods, as well as the residual stresses arising in them.

The reports on various aspects of obtaining fast solidification materials aroused great interest among the Conference participants.

Figure 2. Summing up the results of the 18th International Scientific and Technical Conference Rapid Solidification Materials and Coatings.

Summing up the results of the Conference (Figure 2), the participants noted the high level and wide range of topics of reports on topical issues of the formation by physical and physicochemical methods of surface layers of products with specified properties, the creation of industrial equipment and technologies.

Thus, the 18th International Scientific and Technical Conference Rapid Solidification Materials and Coatings (RSMC–2021) contributed to the further development of this scientific and applied research area and the expansion of professional ties of the participating scientists.

Journal of Physics: Conference Series **2144** (2021) 011001 doi:10.1088/1742-6596/2144/1/011001

Organizing Committee

A A Lozovan, Moscow Aviation Institute (National Research University) – chairman
A V Bespalov, Moscow Aviation Institute (National Research University) – co-chairman
S S Alexandrova, Moscow Aviation Institute (National Research University) – scientific secretary
A M Borisov, Moscow Aviation Institute (National Research University)
D K Kostrin, Saint Petersburg Electrotechnical University
B L Krit, Moscow Aviation Institute (National Research University)
E P Kubatina, Moscow Aviation Institute (National Research University)
I A Petrov, Moscow Aviation Institute (National Research University)

Program Committee

A A Lozovan, Moscow Aviation Institute (National Research University) – chairman
A F Banishev, Institute on Laser and Information Technologies
O A Barabanova, Moscow Aviation Institute (National Research University)
L B Begrambekov, National Research Nuclear University "MEPhI"
S Ya Betsofen, Moscow Aviation Institute (National Research University)
A M Borisov, Moscow Aviation Institute (National Research University)
V A Vasiliev, Moscow Aviation Institute (National Research University)
V A Kazakov, Russian Engineering Union
V P Kazachenko, Belarusian National Technical University
V I Kalita, Institute of Metallurgy and Materials Science
N N Koval, Institute of High Current Electronics
B L Krit, Moscow Aviation Institute (National Research University)
V G Kuznetsov, Institute of Problems of Mechanical Engineering
A E Ligachev, Institute of General Physics
I N Pashkov, Professional Brazing and Soldering Union
L M Petrov, National Institute of Aviation Technologies
M M Serov, Moscow Aviation Institute (National Research University)
V I Shapovalov, Saint Petersburg Electrotechnical University
A V Apelfeld, Moscow Aviation Institute (National Research University)

Peer Review Statement

All papers published in this volume have been reviewed through processes administered by the Editors. Reviews were conducted by expert referees to the professional and scientific standards expected of a proceedings journal published by IOP Publishing.

- **Type of peer review:** Single Anonymous
- **Conference submission management system:** Morressier
- **Number of submissions received:** 37
- **Number of submissions sent for review:** 37
- **Number of submissions accepted:** 37
- **Acceptance Rate (Submissions Accepted / Submissions Received × 100):** 100%
- **Average number of reviews per paper:** 2
- **Total number of reviewers involved:** 22
- **Contact person for queries:**
  Name: Alexandr Alexandrovich Lozovan
  Email: alexandr.lozovan@yandex.ru
  Affiliation: Moscow Aviation Institute (National Research University)

Content from this work may be used under the terms of the Creative Commons Attribution 3.0 licence. Any further distribution of this work must maintain attribution to the author(s) and the title of the work, journal citation and DOI.
Published under licence by IOP Publishing Ltd

The control of the height and shape of the track in laser metal deposition by the QCW laser mode

P S Rodin and V D Dubrov

Institute on Laser and Information Technologies - Branch of the Federal Scientific Research Centre "Crystallography and Photonics" of Russian academy of Sciences, Svyatoozerskaya 1, 140700, Shatura, Moscow Region, Russia

E-mail: vddubrov@mail.ru

Abstract. The control of the track shape in laser metal deposition technology by the QCW laser mode has been studied. The different geometric characteristics of the tracks are shown to obtain at the same average laser power, depending on the selected laser power control mode. The difference in the temperature regimes of track formation is shown.

1. Introduction

The physical processes accompanying the laser metal deposition (LMD) technology are characterized by less stability compared to traditional manufacturing technologies. This variability leads to the appearance of heterogeneities in the construction and limits the use of LMD for the manufacture of critical structural elements, as well as in the aerospace and medical industries [1]. It is necessary to identify anomalies in the process dynamics in a timely and accurate manner to be able to carry out a corrective procedure for ensuring the required level of quality and stability of the LMD process. Thus, it is necessary to control the technological parameters of the LMD in real - time to change the parameters of the applied layer: the temperature and area of the melt region, the height of the applied layer, etc. [2–4]. The development of LMD process control systems is discussed in the review [5]. The powder feed rate is controlled, in particular, to change the height of the track [4]. However, the control system implemented in this work does not allow using the controller developed in existing LMD systems. It is shown in [6] to shift the position of the substrate relative to the nozzle the powder feed rate changes; on the other hand, it is proposed to change the position of the substrate to control the track height. In [7], it is shown how uncontrolled changes in the scanning speed lead to the formation of inhomogeneities in the height of the formed layer. In this paper, we study the influence of pulse-width modulation of laser power on the height and shape of the track in LMD technology.

2. Materials and method

The research facility developed at ILIT RAS was used (figure 1). The setup includes the ytterbium fiber laser (NTO "IRE-Polyus") with a power of 400 W. A robotic arm (Kuka KR10 900-2) moves the substrate relative to the laser head (Precitec YC52). The configuration and speed of the feed disc of the powder feeder (GTV PF 2.1 LC) determine the mass flow rate of the powder. The software part of the management system is implemented using the LabVIEW development environment.

The computing module is equipped with three RS-232 interfaces: one for loading control programs to the CNC controller, and two for switching the serial interface of the CNC communication with the

Content from this work may be used under the terms of the Creative Commons Attribution 3.0 licence. Any further distribution of this work must maintain attribution to the author(s) and the title of the work, journal citation and DOI.
Published under licence by IOP Publishing Ltd

laser "in the gap". Communication with the L-CARD crate, which housed the ADC modules of the multi-channel pyrometer signals, was carried out via the USB interface. Thus, the developed control system intercepts the laser control commands, synchronizes with the radiation on/off commands, and takes into account the current set laser power when generating the corrected values.

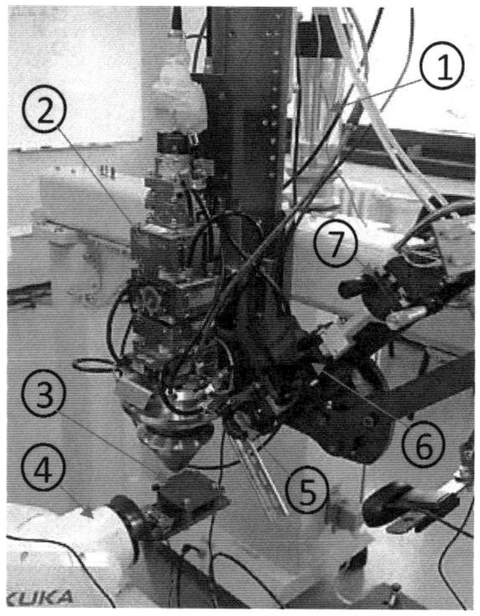

Figure 1. The structure of the experimental setup (the casing is removed): 1 - powder feeder, 2 - laser head, 3 - substrate, 4 - robot manipulator, 5 - pyrometer lens, 6 - pyrometer sensor, 7 - sensor translator.

The control system has access to all aspects of the technological process directly during its execution due to the open architecture of the setup. This allows implementing not only the preliminary planning of the production strategy of the object but also the modification of the strategy "on the go" if such a need is detected as a result of automatic analysis of operational data from the sensors of the installation. Real-time operation imposes increased requirements on the speed and reliability of the control system. To speed up processing, the parallel architecture is used in the implementation of control algorithms.

The track in the LMD process was formed simultaneously with the registration of the optical response. The powder PR-X18N9 with the granulometric composition (40...100) microns was used, which was applied to substrates made of 08X18N10T steel. The powder consumption was 8.4 g/min at the scanning speed of 200 mm/min. During the scanning process, the average radiation power Pa changed stepwise every 12 mm and was equal to 20%, 40%, 60%, 75%, 90% from the nominal laser power. Samples of the results of laser radiation exposure to the substrate are shown in figure 2. The track marked as (1) was made without the use of powder. For tracks (2) and (cw), powder was used. The impact power was changed in cases (1) and (2) with pulse-width modulation, so call QCW mode [8], - with the pulse repetition period t_0=72 ms. In the case of (cw), the power was set at the same levels in continuous mode.

The control system used to control the temperature on the melt surface is the multi-channel optical diagnostic (MCOD) system based on the set of two-color sensors, described earlier in [9]. The radiation of the local areas of the melt is illuminated by separate channels of photosensors. Each photosensor consist of two types of photodiodes mounted on the same optical axis and having sensitivity in different spectral ranges. The spectral sensitivity windows, taking into account the additional optical filters used, are in the ranges (1.7...2.2) microns for the first type of photodiode and (1.1...1.4) microns for the second type of photodiode. Two universal LTR11 ADC modules are used as part of the LTR-EU-8 (L-CARD) crate for digitizing photodiode currents. MCOD system measures the maximum temperature on the surface of the melt bath by the spectral ratio method with a time resolution of about 0.1 ms.

The temperature on the surface of the material at an average laser power P_a of 80 W (20% of the rated power) in continuous mode slightly exceeded the melting point of the material. In the case of using the QCW mode, at the same average power, which corresponds to the fill factor S (the ratio of the pulse duration to the repetition period in the QCW mode) of 20%, - at the irradiation stage, the laser power was 400 W and the value T^*, the maximum temperature on the surface of the melt bath, - exceeded the melting temperature by (400...500) K. QCW radiation is absent in the phase of crystallization.

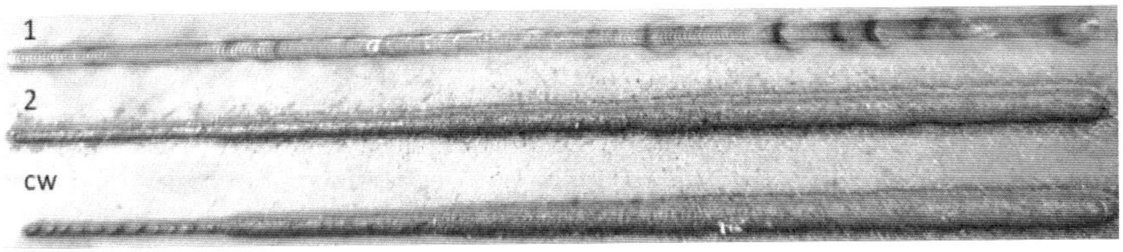

Figure 2. The substrate with single tracks applied. 1. qcw mode of laser power in the absence of powder, period t_0 72 ms; 2. qcw mode of laser power at the powder consumption of 8.4 g/min, period t_0 72 ms; 3. cw – "continuous-wave" mode.

The value of T^* thus at the same average power depends on the mode selection, in the case of (*cw*) with increasing laser power, T^* gradually increases from a temperature just above the melting point to a maximum value of T_m, approximately 2600 K, at a laser power of 400 W. In mode (2), heating occurs during the period of laser action at the nominal laser power level P_0: $P_a = S^* P_0$. As S increases in QCW mode, the time of temperature drop at the stage of no laser radiation generation (1-S) t_0 decreases, the t_0 increases while heating duration S, and thus the value T^* increases to the same value T_m.

3. Results and discussion

Dependences of the melt bath of width D, height H, and the shape coefficient R [ratio of height to width] of the track are shown on figure 3 on the maximum surface temperature T^* at the speed of 200 mm/min and the powder consumption of 8.4 g/min. As follows from the results obtained, figure 3, it is possible to change the track height in the wide range by adjusting the average power id est by changing S in QCW mode. The track width in both modes increases with increasing temperature. In the (*cw*) mode, the track height is weakly dependent on the laser power. As the result, it is possible to change the track shape factor from 0.4 to 1.6 depending on the selected mode and the average power level. In [3–5], data on the stabilization of the melt temperature by controlling the laser power are presented. It is shown in [6–7] that the ultimate goal of the control system is to stabilize the parameters of the formed layer, first of all, the track height, otherwise, on the next layer, the change in the distance from the layer surface to the nozzle leads to defects in the construction of the next layer. Using QCW mode allows you to control the track height by changing the S.

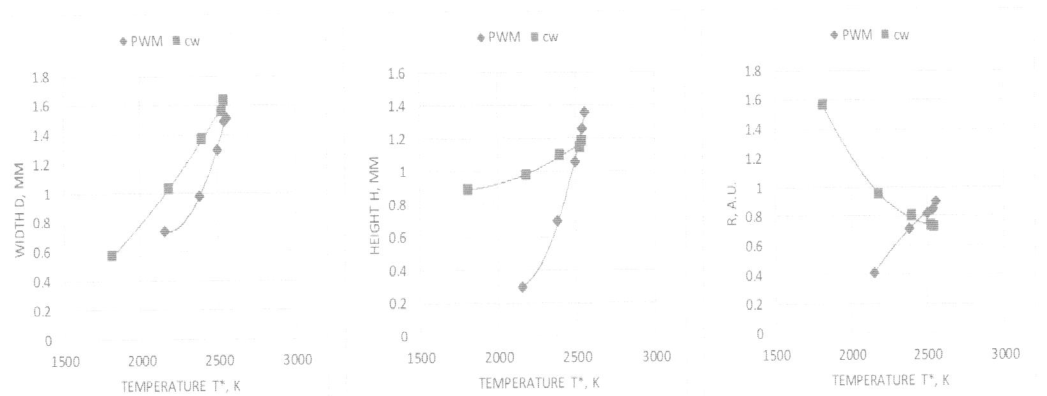

Figure 3. The dependence of the width D, height H and the shape coefficient R on T^* at a speed of 200 mm per minute and at a powder consumption of 8.4 g/min.

The estimate of the cooling rate as the function of S obtained by processing the time sequence of data from the pyrometric unit is shown in table 1. The cooling rate of the melt surface is higher with the increase in the duration of the pause between the radiation pulses in the QCW mode, and therefore with the increase in *S*. Thus, in QCW mode, the cooling rate can be adjusted by changing the duty cycle. The increase in the cooling rate to 10^5 K/s and higher leads to additional effects in the formation of the microstructure in chromium-nickel steel [10].

Table 1. The cooling rate.

	S, %	Cooling rate, K/s
1	20	$0.12*10^6$
2	40	$0.08*10^6$
3	60	$0.014*10^6$

4. Conclusions

Thus, the setup developed at ILIT RAS with the open control system architecture. It is used to study the effect of pulse-width modulation of laser power on the width and shape of the track in the LMD technology. The control system is placed "in the gap" between the existing CNC controller and the laser to allow the integration into existing systems, and the use of pulse-width modulation for adjusting the height of the track. PID control based on the signal of mismatch of the set temperature with the data of the multichannel pyrometer can be enabled separately or, if necessary, in conjunction with QCW mode of the laser power. The dependences of the width, height, and shape coefficient on the maximum temperature on the melt bath surface for two laser operating modes, cw and QCW, are obtained. It is shown to change the track shape coefficient from 0.4 to 1.6 depending on the selected mode and the average power level. In addition, it is possible to change the track height in a wide range by adjusting the power fill factor in QCW mode. The width of the track in both modes increases equally with increasing temperature. The track height is weakly dependent in cw mode on the laser power. The cooling rate in the PWM mode at the crystallization stage is estimated from the dynamics of temperature changes on the melt surface. It is shown to be possible in the QCW mode the regulation of the cooling rate at the crystallization front by changing the duty cycle. The obtained results are applicable in the LMD technology for controlling the geometric and microstructural parameters of the formed track.

Acknowledgements

The work was supported by the Ministry of Science and Higher Education of the Russian Federation within the framework of the State Task of the Federal State Budgetary Institution “Crystallography and Photonics” of the Russian Academy of Sciences in terms of the developed methodology, as well as with the grant support of the Russian Foundation for Basic Research (Project No. 18-29-03249) in terms of conducting experiments and the experimental results obtained.

The authors are grateful to Dr. Y N Zavalov (Institute on Laser and Information Technologies - Branch of the Federal Scientific Research Centre "Crystallography and Photonics" of Russian academy of Sciences) for his guidance and support.

References

[1] Mazumder J 2017 *Laser additive manufacturing (Woodhead publishing series in electronic and optical materials),* ed M Brandt (Woodhead Publishing) 21–53
[2] Turichin G, Somonov V V, Babkin K D and Klimova O G 2015 *Physics Procedia* **78** 397
[3] Bi G, Gasser A, Wissenbach K, Drenker A and Poprawe R 2006 *Optics and lasers in engineering* **44(12)** 1348
[4] Akbari M and Kovacevic R 2019 *The Int. J. of Adv. Manuf. Technol.* **104(5)** 2887
[5] Shamsaei N, Yadollahi A, Bian L and Thompson S M 2015 *Additive Manufacturing* **8** 12

[6] Zhong C, Pirch N, Gasser A, Poprawe R and Schleifenbaum J H 2017 *Metals* **7(10)** 443
[7] Tang Z J, Liu W W, Wang Y W, Saleheen K M, Liu Z C, Peng S T, Zhang Z and Zhang H C 2020 *The Int. J. of Adv. Manuf. Technol.* **108** 3437
[8] Xiao H, Li S, Han X, Mazumder J and Song L 2017 *Materials & Design* **122** 330–9
[9] Zavalov Y N, Dubrov A V, Rodin P S and Dubrov V D 2018 *Proc. SPIE* **10675** 1067511
[10] Xiao H, Li S M, Xiao W J, Li Y Q, Cha L M, Mazumder J and Song L J 2017 *Materials Letters* **188** 260

Obtaining and plasma-electrolyte modification of fibers of ultralight magnesium alloy

B L Krit[1,2], M M Serov[1], N V Morozova[2,3], Ruizhi Wu[4] and V M Medvetskova[1]

[1] Moscow Aviation Institute (National Research University), 125993, Moscow, Russia
[2] Moscow State University of Technology «STANKIN», 127055, Moscow, Russia
[3] Russian Medical Continuous Professional Education Academy, 123995, Moscow, Russia
[4] Harbin Engineering University, Harbin 150001, P.R. China

E-mail: bkrit@mail.ru

Abstract. The results of researches for the transformation to fiber state the Mg-8Li-1Al-0.6Ce-0.3Y ultralight magnesium alloy by the Pendant Drop Melt Extraction (PDME), and subsequent modifying obtained fibers by Plasma-Electrolytic Treatment (PET) are presented. The results demonstrate possibility of successful application of the above-mentioned methods in relation to chemically active materials. Purposeful modifying of ultralight magnesium alloys by PDME and PET methods is capable to significantly expand areas of using to these materials.

1. Introduction

Fiber materials occupy a special place in modern applied materials science and are used in various branch of industries [1–3]. Metal fibers are obtained by mechanical and physicochemical methods, as well as by crystallization from melts [4]. A promising technological process for the production of fibers is the method of Pendant Drop Melt Extraction (PDME) [5, 6], which provides high crystallization rates. The main advantage of the PDME method over other casting methods for producing fibers is the use of crucible-free melting, which makes it possible to obtain fibers from chemically active and refractory metals (including nickel, titanium, zirconium, etc.), heat-resistant, corrosion-resistant and hard-to-deformation steels and alloys.

Studies of magnesium alloy fibers obtained by rapid crystallization methods are actively conducted in the Russian Federation and abroad. In [7] provides a detailed analysis of various methods for rapid crystallization of magnesium alloys. It is noted, that an increasing in the cooling rate of the metal melt can cause cardinal structural changes in alloys up to an amorphous state. For magnesium alloys, the technology of fast crystallization with cooling rates of the order of 10^6-10^8 K/s is very promising. Depending on the value of the cooling speed, the design of the plants, the influence of additional technological factors, structures are formed that have a significant impact on the properties of both the fibers themselves and the materials made from them. Ultra-fast crystallization contributes to the grinding of structural components: dendritic cells and crystals of secondary phases. Supersaturated solid solutions and metastable phases are formed.

Content from this work may be used under the terms of the Creative Commons Attribution 3.0 licence. Any further distribution of this work must maintain attribution to the author(s) and the title of the work, journal citation and DOI.
Published under licence by IOP Publishing Ltd

With relate to magnesium alloys, it has been shown that high-speed solidification of the melt significantly improves the uniformity of the structure, which leads to increased strength and ductility, regulates corrosion resistance and expands the possibility of their use in the aviation industry, mechanical engineering, medical engineering and many other industries. It should be noted that the application of the fast crystallization method for magnesium alloys, which include light elements (for example, Li and Be), rare earth or IIIA transition metals (Sc, Y, La), is poorly understood. At the same time, in a number of works there is a special interest and prospect of using high crystallization rates precisely for magnesium alloys of similar doping systems [7–9].

To give functional properties, as well as to protect against external influences (which is especially relevant for fibers made of light and ultra-light alloys), modifying layers and coatings on the surface need formed. All of the above structural-phase transformations due to the production at high-speed crystallization fibers are likely to affect the processing and conversion of the fiber state alloy, in particular – the plasma-electrolyte modification process [10]. The present study is devoted to the study of these features.

2. Materials and methods

In this work, to obtain fibers we used an ultralight alloy with the composition Mg-8Li-1Al-0.6Ce-0.3Y prepared at the Key Laboratory of Superlight Materials & Surface Technology Ministry of Education, Harbin Engineering University. The alloy manufacturing technology included melting of ingots of commercially pure Mg (> 99.99 wt%), Li (> 99.99 wt%) and Mg-20 (Ce-Y) master alloy in a medium-frequency vacuum induction furnace in an argon atmosphere at 720 °C during 20 min. The melt was poured into a permanent mold (130 mm × 110 mm × 40 mm) at a temperature of 700 °C. Then the as-cast alloy was cut into plate-shaped blanks with dimensions of 130 × 30 × 3 mm.

Figure 1 shows a diagram of the fibers production by the Pendant Drop Melt Extraction (PDME) method.

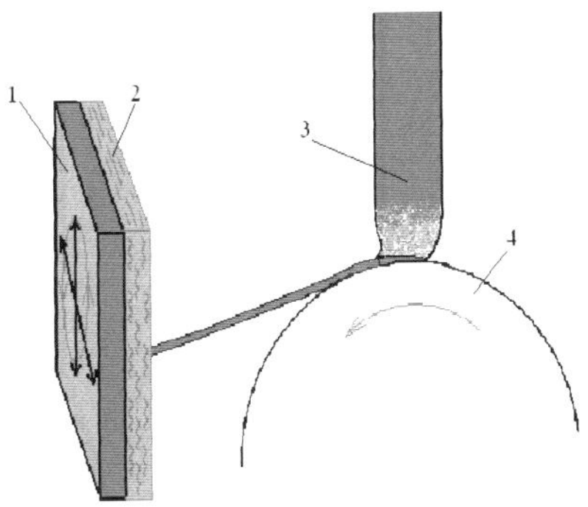

Figure 1. PDME scheme: 1 – receiving plate; 2 – canvas made of alloy's fibers; 3 – workpiece with a drop of melt; 4 – rotating cooled disc.

A graphite heater to melt the workpiece' butt and form a hanging drop of the melt was used. The design of the heater allows the process of drawing fibers simultaneously from several workpieces. Advantages of the PDME is application of crucible-free melting in a protective atmosphere that does possible producing the fibers even from refractory and chemically active metals. Extraction of the fibers from melt by PDME method occurs at cooling rate up to 10^6 K/s. The method allows to form both lengthy fibers, and discrete particles and also forming porous nonwoven canvases.

The conversion of the blank into fibers is as follows. The lower butt of the vertically located workpiece melts and form a pendant drop, with which the top of the working edge of the rotating cooled disk (heat receiver) contacts. Local solidification of melt occurs in contact zone. Due to the rotation of the heat receiver, the solidified material is escaped out of the melt in the form of a fiber and, under the influence of centrifugal forces, is ejected from the top of the working edge. During operation, as the melt is spending, the workpiece is lowered into the melting zone and, thus, the volume of the melt drop is replenished. By controlling such process parameters as the rotation speed of the heat receiver, the melt feed rate and the heating power, it is possible to obtain a fiber material in the form of a "cotton wool" (with an individual fiber length of more than 10 mm and an equivalent diameter of 30-80 microns).

Before PET, fibers were degreased in ethanol and then dried in the open air during 5 min. The PET treatment technique and the equipment used are described in [10, 11]. The PET process was carried out on the capacitor-type equipment in the anodic-cathodic mode at the different current densities in the voltage range of 300–500 V and the currents ratio I_{an}/I_{cat} range of 0.8–1.2. The electrolyte temperature was maintained at the level of 20–25 ^{0}C.

Electron microscopy and EDS researches were carried out on the Quanta SEM analysis system with application of the Genesis Software program.

3. Results and discussion

A typical photograph of an individual fiber is shown in figure 2, the elemental analysis results are shown in figure 3.

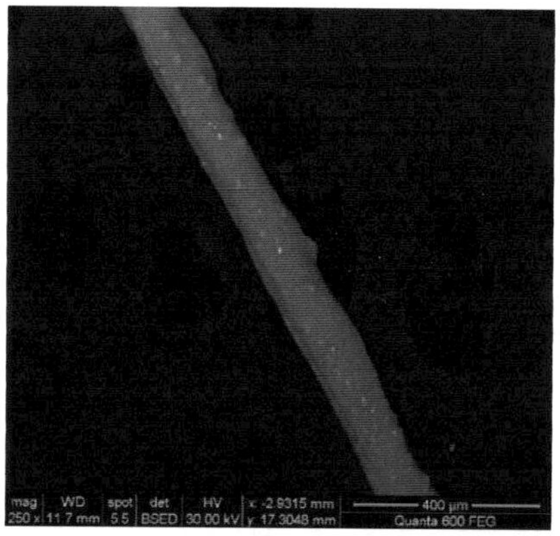

Figure 2. The appearance of the Mg-8Li-1Al-0.6Ce-0.3Y alloy fiber.

Given the high pyrophoricity of the magnesium alloy, the process was carried out in a protective atmosphere, for which the working chamber of the installation was filled with argon at a pressure of ≈ 80 kPa. The obtained fibers in section have a sickle shape what is associated with the configuration of the contact edge of the heat receiver. Thickness of fibers changed within 50 ÷ 70 microns, width of 90 ÷ 150 microns, length of 30 ÷ 120 mm. The elemental composition of the alloy after conversion into the fiber state has not changed significantly (figure 3).

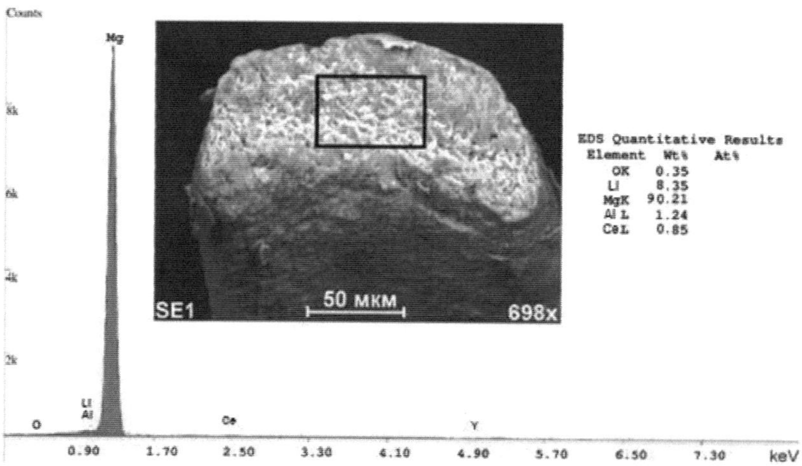

Figure 3. Cross-sectional view and EDS fiber analysis data.

Attempts of plasma-electrolyte treatment of fibers confirmed the presence of problems noted in our previously studies [11]. Using the electrolyte and the mode we tested earlier for ML10 alloy did not give an acceptable result. The fibers ignited and burned at the initial stages of initiation of the micro-discharges. A similar situation was observed when using an electrolyte of the composition 10 mL/L $(Na_2O)_m(SiO_2)_n$ + 3 g/L KOH + 4 g/L KF $2H_2O$ which showed its operability when treating a compact alloy Mg-8Li-1Al-0,6Ce [11].

To the encourage result was led by the treatment in electrolyte 1.2 g/L KOH + 15 g/L $NaAlO_2$ [11], but at the same time a softer electric mode was required. A more or less continuous coating with a thickness in range of 1-2 μm on the Mg-8Li-1Al-0.6Ce-0.3Y alloy were getting to form during processing for 5 minutes in the aluminum-alkaline electrolyte. The anodic-cathodic mode at the current density of 5 A/dm^2, the voltage of 350 V and the currents ratio I_{an}/I_{cat} = 1.1 were applied. The typical appearance and cross section of the fiber after the PET is shown in figure 4.

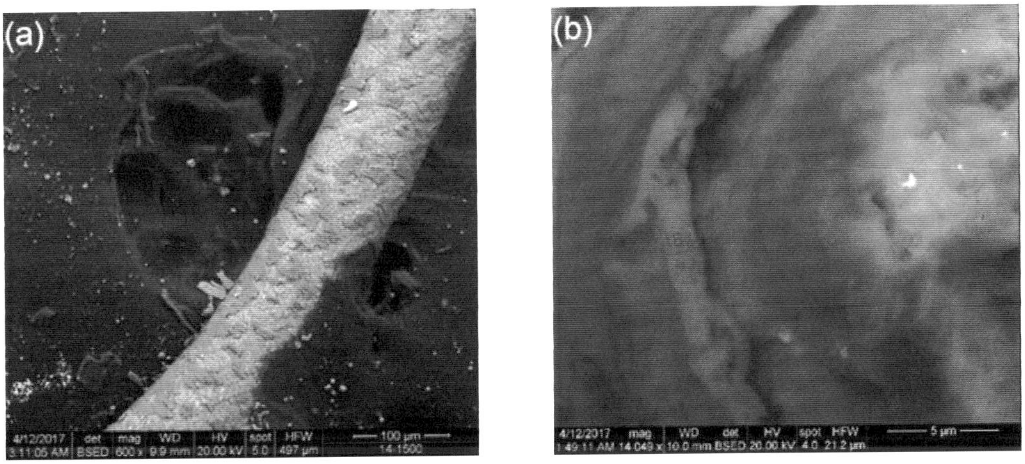

Figure 4. The appearance of the fiber (a) and its cross section (b) after the PET.

As a result of executed researches, the modes of obtaining and the subsequent plasma and electrolytic modifying of fibers of ultralight magnesium Mg-8Li-1Al-0.6Ce-0.3Y alloy were found. It should be noted that that PET process for the ultralight alloy transformed in fibers by PDME method is realized more difficult than processing fibers of ML10 alloy or the same alloy in a compact state. Nevertheless, appropriate regimes have been found that have made it possible to achieve the goals set.

The conversion of the compact alloy into the fibers form did not entail any noticeable change in composition. As well it is known that thin coatings formation can improve the characteristics of materials without significantly changing in their shape and dimensions [12, 13]. Thus, joint application of the PDME and PET methods for purposeful modifying is capable to significantly expand possibilities of magnesium alloys using.

4. Conclusions

1. Experimentally determined the conditions for the implementation of PDME for the production of fibers from a chemically active ultra-light alloy based on magnesium.

2. It has been found that as a result of transformation into a fiber form, the alloy retains its element composition.

3. he possibility of plasma-electrolyte treatment for fibers of the ultralight alloy based on magnesium is shown.

Acknowledgments

This work was carried out with the financial support of the Russian Foundation for Basic Research in scientific project No. 20-58-53023 \ 21.

References

[1] Pelevin F V 2007 *Bulletin of the Association of Universities of Tourism and Service* **1(3)** 46–51
[2] Kostornov A G 1983 *Permeable metal fiber materials* (Kiev: Technique) 128 p
[3] Farafonov D P, Migunov V P, Degovets M L and Alyoshina R S 2016 *Electronic scientific journal "Proceedings of VIAM"* **4** http://viam-works.ru/ru/articles?art_id=941
[4] Belov S V, Vityaz P A, Sheleg V K et al. 1987 *Porous permeable materials* (Moscow: Metallurgy) 335 p
[5] Shevakin Yu F and Eskin T I 1998 *Fundamental problems of the Russian metallurgy at the threshold of the 21st century* **2** 486
[6] Farafonov D P, Degovets M L and Aleshina RSh 2016 *Aviation materials and technologies* **40(1)** 44–52
[7] Volkova E F 2006 *MITOM* **11** 5–9
[8] Ilyin A A, Fatkullin O Kh, Stroganov G B, Shulga A V and Martynov V N 2008 *Structure and properties of fast-hardened alloys* (Moscow: Altex) 588 p
[9] Tao W, Milin Zh, Zhongyi N and Bin L 2006 *J. of Rare Earth* **24(6)** 797–800 https://doi.org/10.1016/S1002-0721(07)60032-5
[10] Rakoch A G, Dub A V and Gladkova A A 2012 *Anodization of light alloys under various electric modes. Plasma-electrolytic nanotechnology* (Moscow: Old Basmannaya) 496 p
[11] Grigoriev S, Kondratsky I, Krit B et al. 2021 *J. Eng. Mater. Technol.* 1–15 https://doi.org/10.1115/1.4052718
[12] Vereschaka A, Tabakov V, Grigoriev S, Sitnikov N, Andreev N and Milovich F 2018 *Wear* **416-417** 72–80 https://doi.org/10.1016/j.wear.2018.10.004
[13] Grigoriev S, Peretyagin P, Smirnov A, Solis W, Diaz L A, Fernandez A and Torrecillas R 2017 *J. Eur. Ceram. Soc.* **37** 2473–9 https://doi.org/10.1016/j.jeurceramsoc.2017.01.027

Peculiarities of calculations of formation of extended sections of steel castings in disposable moulds with functional coating of definite microrelief

A F Smikov[1], S S Kuzovov[2] and V S Moiseyev[1]

[1] Moscow Aviation Institute (National Research University), 121552, Russia, Moscow, Orshanskaya str., 3
[2] FSBEI HE «Bryansk State Technical University» 241035, Russia, Bryansk, 50 Let Oktyabrya, 7

E-mail: kss41188@inbox.ru

Abstract. In the present article are considered the issues of influence of the microrelief of the working cavity surface of the sand-and-clay mould on the intensity of heat exchange and formation of the structure in the surface area of the metal blank, for prevention of hot cracks formation.

The development of technological means for struggle against hot cracks in steel castings obtained in disposable, including sand-and-clay, moulds remains actual at the present time [1–5]. As one of the measures for preventing such defects during formation of steel castings, it is possible to use a special relief coating 4-5 mm thick of the inside surface of the mould. Such coating, as the research shows, creates additional obstacles to formation of hot cracks (HC) on account of formation in the surface zone of a solidifying alloy of stronger structure - a "reinforced" layer, dispersal of stresses in the casting in the period of its crystallization, as well as reduces heat expansion of the mould base.

During making a definite microrelief on the surface of the mould working cavity the chromite sand of fractions 0,2 and 0,4 has been suggested. The use of a functional coating based on the chromite sand of definite fractions forms a microrelief on the surface of the casting mould working cavity, which changes conditions of its heat exchange with the melt of the alloy. As a result, morphology of dendrite crystals in the surface zone of the casting is changed. In this connection, it is suggested to use the method of calculation of intensity of heat exchange of the metal blank being formed, with a combined shape, which allows to determine the value of the effective coefficient of heat accumulation of the mould with a functional coating, during solidification of the steel casting. This method is also used in the calculations of duration of solidification of steel castings sections, and their linear rates of crystallization.

The extent of the coating relief is considered by the coefficient of the mould $К_f$, which is determined from the expression:

$$К_f = \frac{S_{cov}}{S_{cov_0}}, \quad (1)$$

where S_{cov} - actual surface area, S_{cov_0} - geometrical surface area, m^2.

Content from this work may be used under the terms of the Creative Commons Attribution 3.0 licence. Any further distribution of this work must maintain attribution to the author(s) and the title of the work, journal citation and DOI.
Published under licence by IOP Publishing Ltd

The actual surface area of the casting mould working cavity with a functional coating was determined by a 3d model of this surface, obtained by the method of volume reconstruction. Two layers of the casting mould are suggested to be considered on account of application of the effective coefficient of the mould heat accumulation b_{ef} [6]. Considering a relief surface of the mould the following expression was received for calculation of the value b_{ef}, and adeq1uacy of it was checked in CSS "PoligonSoft":

$$b_{ef} = \left\{1 + 2\sum_{n=1}^{\infty}(-1)^n \left(\frac{1-K_b}{1+K_b}\right)^n exp\left(\frac{-n^2}{Fo}\right)\right\} b_{cov}, \quad (2)$$

where $K_b = \frac{b_m}{b_{cov}}$; $Fo = \frac{a_{cov}\tau}{\delta'^2_{cov}}$, $\delta'_{cov} = K_f\delta_{cov}$; b_{ef}- efficient coefficient of the mould heat accumulation, $W \cdot s^{1/2}/(M^2 \cdot K)$; b_m, b_{cov} – coefficient of heat accumulation of the mould and coating, accordingly, a_{cov} – thermal diffusivity of the coating, m^2/s; τ – time, s; δ_{cov} - coating thickness, m.

According to expression (2) b_{ef} is the function of time. Its magnitude, dependent on the value of Fourier criterion Fo, is in the limits from b_{cov} to b_m. Calculation of time of the casting section solidification τ_s is determined from the equation of its heat balance by the iteration method, because it is a function of the effective coefficient of the combined mould heat accumulation b_{ef} and its initial temperature. Thus, for extended elements of the casting τ_s blank is calculated according to the formula:

$$\tau_s = \left(\frac{C}{B}\right)^2 /\psi, \quad (3)$$

where $C = \rho q_{eq}\frac{X_c}{K_f}$ and $B = \frac{2}{\sqrt{\pi}} b_{ef}(T_p - T_{m.i})$ – complexes in the equation of heat balance, considering K_f; ψ - calculated coefficient considering irregularity of a temperature field along the mould section during its solidification; ρ - density of the alloy, kg/m^3; X_c - adjusted dimension, m; T_p - mean calculated temperature of the casting in the process of its solidification, K; $T_{m.i}$ - initial temperature of the mould, K.

Investigations of the value of the mould heat accumulation coefficient with a functional coating based on the chromite sand, from the relationship of the coating thickness to the adjusted dimension of the casting, with consideration of the mould inside surface relief (figure 1) showed that the value b_{ef} increases with the rise of relationship of the coating thickness to the thickness of the casting wall .

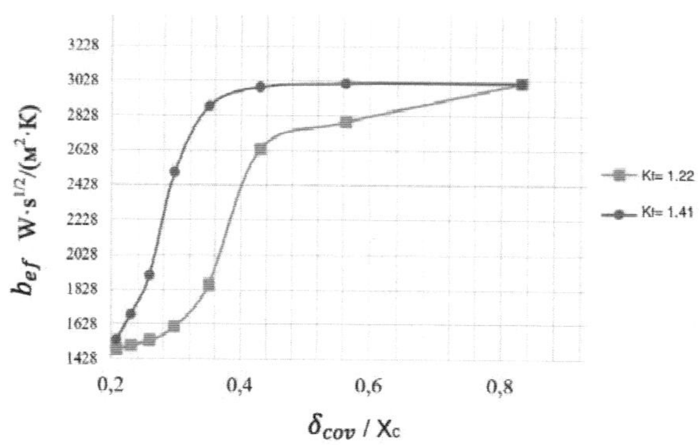

Figure 1. Dependence of the value b_{ef} of the mould from relationship of coating thickness to adjusted dimension of the casting with consideration of K_f.

From the graph in figure 1 also follows that coarsening of the chromite sand fraction for the functional coating results in a quicker growth of b_{ef}, other things being equal, and can serve for explanation of mechanism which is observed in the photos of microstructure of A352GrLCC steel (figure 2) crystallized at the surface of the casting, with different values of microrelief.

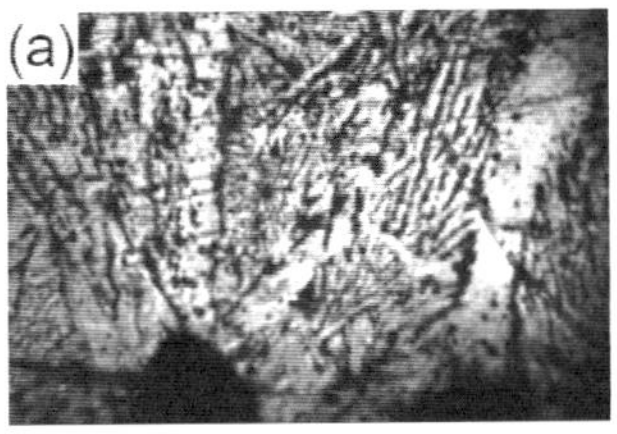

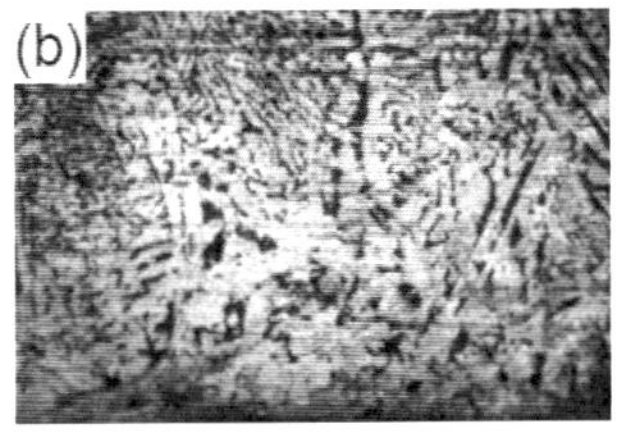

Figure 2. Microstructures of A352GrLCC steel in the wall area, with different microrelief of coating, x 100: a) K_f = 1.22; b) K_f = 1.41.

Namely, with crystallization on the mould surface with value K_f = 1.22 the mechanism of heterogeneous nucleus formation is realized directly on the peaks of chromite sand. The structure of the wall area of the mould, received on the surface with K_f = 1.41, in the area directly adjoining to the casting mould, is characterized by the presence of the zone of small "frozen" crystals, which goes to the zone of dendrite multidirectional crystals under the influence of temperature front due to the presence of microrelief.

According to [7], the rate of linear crystallization of steel in the sand-and-clay mould is found from the expression:

$$V_k = \frac{k_0^2\left(1+\frac{D}{k_0^2}\right)^2}{2(x+0{,}002\ \Delta T_M X_c)}, \quad (4)$$

where k_0 – calculated coefficient of volume solidification; ΔT_M – overheating of steel higher than temperature of solidification, K; D – a constant, x – distance from the mould surface, m.

The coefficient of volume solidification k_o is the function of time of solidification of the mould section being considered τ_S, and of its adjusted dimension X_c, and can be found from the expression:

$$k_0 = \frac{X_c}{\sqrt{\tau_S}}, \quad (5)$$

In the result of the fulfilled calculations the curves of linear rate of crystallization have been plotted for different values K_f dependent on the distance to the boundary of the casting mould (figure 3).

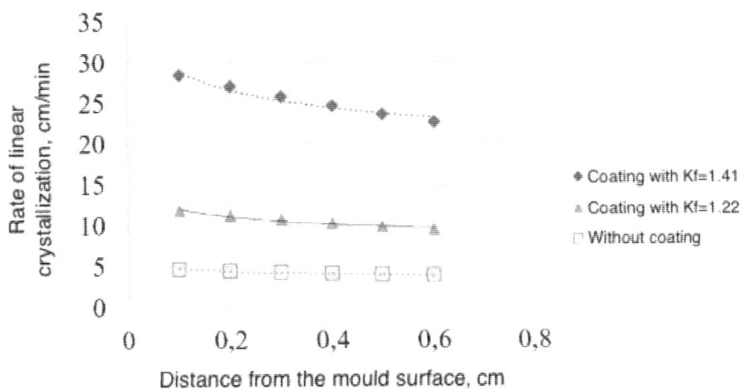

Figure 3. Change of rate of linear crystallization at different distances from the mould surface and for different surfaces.

It is known [7] that the increase of linear crystallization rate results in the increase of the volume crystallization rate, which, in turn, is confirmed with the results of microstructural analysis of the mould surface zone, characteristic for the functional chromite coating based on fraction 0,4 mm. The received curves, according to the known relationships between the crystallization rate and the inter-axial distance of the dendrite structure [8], allow to evaluate dispersivity of the structure and to predict the level of mechanical properties of the melt, based on the mechanisms presented in works [9–11].

In figure 4 are presented photos of macrostructure of the mould sections subjected to formation of the defect HC (hot crack), made by different technologies. White lines drawn on the photos allow to trace the directions of growth of main axes of dendrits.

Figure 4. Photos of macrostructure of specimens from steel A352GrLCC: a) casting with HC, made by the usual technology; b) – casting without HC made with application of functional coating.

In castings made without application of the functional coating (figure 4, a), zone of columnar dendrites, main axes of which are oriented by the normal to the surface, adjoins directly to the mould surface. Such structure presented in the surface layer of the casting does not provide a sufficient level of mechanical properties in the temperature interval of crystallization, in order to resist the formation of HC. The application of a functional coating (figure 4, b), creates the microrelief on the casting mould surface which contributes to deviation of the heat flow direction from the normal to the surface of contact. This changes the configuration of the temperature field, which, in the vicinity of the division surface, repeats the profile of microrelief of the casting mould surface, which leads to the significantly more expressed manifestation of the mechanism of thinning of dendrites in the surface zone of the casting. As a result, a layer of multidirectional crystals of the casting surface zone is formed, which increases the strength of metal in solid-and-liquid condition, and in the process of solidification it prevents formation of surface HC.

References

[1] Korotchenko A Yu 2011 *Liteishik Rossii* **12** 17–21
[2] Vdovin K N, Savinov A S and Feoktistov N A 2014 *Liteynoye proizvodstvo* **12** 8–10
[3] Larichev N S and Korotchenko A Yu 2020 *Zagotovitelnie proizvodstva v mashinosnroyenii* **18** 99–9
[4] Dantzig J A and Rappaz M 2009 *Solidification* 644 p
[5] Drezet J M, Gremaud M, Graf R and Gaiimann M 2002 *Proceedings of the 4th European Continuous Casting Conference, IOM communications, Birmingham* 755–63
[6] Neustruyev A A, Moiseyev V S and Smikov A F 2005 *Razrabotka SAPR tehnologicheskikh protsessov litya* 216 p
[7] Hvorinov N I 1958 *Kristallizatsiya I neodnorodnost stali* 392 p
[8] Gulyayev B B 1976 *Teoriya liteinikh protsessov* 211 p
[9] Golod V M and Denisov V A 2007 *Teoriya, kompyuterniy analiz I tehnologiya stalnogo litya* 610 p
[10] Rappaz M, Grasso P D, Mathier V, Drezet J M and Jacot A 2004 *The Minerals, Metals & Materials Society*
[11] Flemengs M S 1977 *Protsessi zatverdevaniya* 423 p

Production and investigation of powders for additive synthesis from a new beta-solidifying TiAl-based alloy

P V Panin[1], I A Bogachev[1] and E A Lukina[2]

[1] Titanium Alloys Lab., Federal State Unitary Enterprise "All-Russian Scientific Research Institute of Aviation Materials", 17 Radio Str, 105005, Moscow, Russia
[2] Department of Materials Science and Technologies of Materials Treatment, Moscow Aviation Institute (National Research University), 4 Volokolamskoe Shosse, 125993, Moscow, Russia

E-mail: PaninPaV@yandex.ru

Abstract. Chemical composition, structure, and technological properties have been investigated for metal powder compositions (MPCs) of a new six-component TiAl-based alloy with Gd microadditions: Ti-31.0Al-2.5V-2.5Nb-2.5Cr-0.4Gd, wt.% (Ti-44.5Al-2V-1Nb-2Cr-0.1Gd, at.%). Three MPCs fractions (10–63, 40–100, 80–120 μm) were produced by electrode induction melting and inert gas atomization technique and targeted for the additive synthesis of parts. It is shown that the chemical composition of the MPCs for the main elements corresponds to that of the electrode. In contrast, a 1.5-fold increase of the oxygen content in the MPCs was observed, which is being the result of natural oxidation of powder particles upon air environment due to developed specific surface. It has been determined that the phase composition of the MPCs ($\gamma+\alpha(\alpha_2)+\beta$) differs from the equilibrium phase composition of the electrode ($\gamma+\alpha_2+\beta_0$/B2) and corresponds to a rapidly quenched metastable state, which indicates high solidification rates in the atomization process, exceeding critical cooling rates of the alloy. The technological properties, specifically the powder flowability, were found to be improved for 40–100 and 80–120 μm fractions, making them applicable for additive synthesis of parts from the studied alloy by selective electron-beam melting method.

1. Introduction

Metal powders intended for additive synthesis, obtained by atomization of bulk workpieces (electrodes), can be considered as rapidly crystallized *in-situ* granules [1, 2]. To date, a large number of successful studies have been carried out on traditional titanium alloys (primarily Ti-6Al-4V) obtained by various additive technologies. Meanwhile, the volume of research is insufficient for intermetallic alloys based on γ-titanium aluminide TiAl – *gammalloys*, which are promising for high-temperature applications [3–7]. The majority of studies were carried out on the 2nd generation of TiAl-based alloys containing 47–48 at.% Al and solidifying via the peritectic reaction $L+\beta\rightarrow\alpha$. Moreover, more than 80 percent of the studies are devoted to the Ti-48-2-2 alloy – the only one commercially used (in GEnx engines for the Boeing 787 Dreamliner) [7]. Fewer works are in the field of atomization and synthesis of TiAl-based alloys of the 3rd and subsequent generations which solidify entirely through the β phase field – nevertheless, in the last several years this topic has acquired intensive development at the Montanuniversität Leoben (Austria) by one of the world's

Content from this work may be used under the terms of the Creative Commons Attribution 3.0 licence. Any further distribution of this work must maintain attribution to the author(s) and the title of the work, journal citation and DOI.
Published under licence by IOP Publishing Ltd

famous groups in the field of gamma alloys leading by H. Clemens and S. Mayer in cooperation with MTU Aero Engines AG [8–11].

FSUE "VIAM" also conducts systematic research in this direction. The object of the present study is a new Russian intermetallic β-solidifying TiAl-based alloy of the following nominal composition: Ti-31.0Al-2.5V-2.5Nb-2.5Cr-0.4Gd, wt.% (Ti-44.5Al-2V-1Nb-2Cr-0.1Gd, at.%). The alloy was developed at FSUE "VIAM" [12]. The aim of the work was to investigate the chemical composition, structure, and technological parameters of various size fractions of metal powder compositions (MPCs) produced from the studied alloy.

2. Materials and methods

The initial charge cylindrical billet (electrode 80 mm in diameter) from the investigated TiAl-alloy was produced by a combined vacuum-arc (VAR) and vacuum induction melting (VIM) method with triple remelting (VAR + VIM + VAR) in ALD VAR L200 and ALD Leicomelt 5 furnaces. The MPCs have been obtained by crucible-free electrode atomization on a VIPiGR 50/500 vacuum induction melting and gas atomization unit designed and manufactured at FSUE "VIAM". The cone-shaped working edge of the rotating electrode is gradually melted in an annular inductor, and quasi-continuously flowing melt droplets fly through a special nozzle and dispersed by high-pressure argon, thus crystallizing into powder particles (EIGA technology – *electrode induction melting and inert gas atomization* [13]). During the spraying process, the electrode and the melt do not come into contact with any surface which excludes contamination of the resulting powder. The temperatures of critical points which are necessary for the rational choice of the atomization parameters were determined for the studied alloy in our previous research ($T_{\text{liquidus/solidus}} \approx 1530 / 1470$ °C) [14].

The powder extracted from the receiving hopper of the atomizer was subjected to separation on a SWECO vibrotrunk to remove scrap and powder particles with a size of more than 160 μm; the fraction less than 10 μm was separated on a gas-dynamic classifier of a HERMIGA 10/100 VI unit. The particle size distribution was determined using laser particle size analyzers Mastersizer 3000E (Malvern PANalytical) and Analyzette 22 NanoTech. Three main fractions were obtained by means of dry screening in an A20/4 sieve analyzer: 10–63, 40–100, and 80–120 μm.

The MPCs' chemical composition was investigated on a Bruker S8 Tiger X-ray spectrometer. The shape and structure of particles were studied with a Jeol JSM-6490LV scanning electron microscope (SEM). X-ray diffraction analysis (XRD) was carried out at room temperature on a Bruker D8 ADVANCE diffractometer in CuK_{α} radiation. The experimental diffraction patterns were processed using DIFFRAC.SUITE and TOPAS software with a licensed copy of JCPDS-PDF-2 Database.

3. Results and discussion

For the production of MPCs, several technologies are known, which are divided into two large groups. The first group includes technologies with crucible melt production (HERMIGA type). For this purpose, ceramic crucibles are used in which charge mixtures of grade composition of alloys are melted. The second group includes crucible-free technologies implemented in a protective atmosphere and assuming not a simultaneous melting of the entire initial charge, but the gradual melting of a compacted cylindrical workpiece (electrode). Crucible-free technology is preferable for high reactive alloys, primarily based on titanium and its intermetallic compounds, thus it was used in our work.

The particle morphology was assessed for all investigated MPC fractions. It has been established that the main defects of the obtained powders are "satellites", being the result of both the difference in the flight speed of different size particles, and the ingress of small particles into the primary melt splitting zone with turbulence formed during atomization. Particles-"satellites" in an amount of up to 10 vol.% are typical defects of metal powders obtained by gas atomization process, which is not a critical deviation from the norm for intermetallic alloys (figure 1a). All powder particles have a dendritic structure (figure 1b) which is common for the pre-solidus quenched state as a result of rapid cooling of melt droplets when spraying with a gas (argon).

The analysis of the powder particles sphericity was carried out according to two criteria: (1) assessment of the elongation of a particle, and (2) assessment of the shape by the ratio of the perimeter of a particle (P) to the square root of its area (S) (table 1).

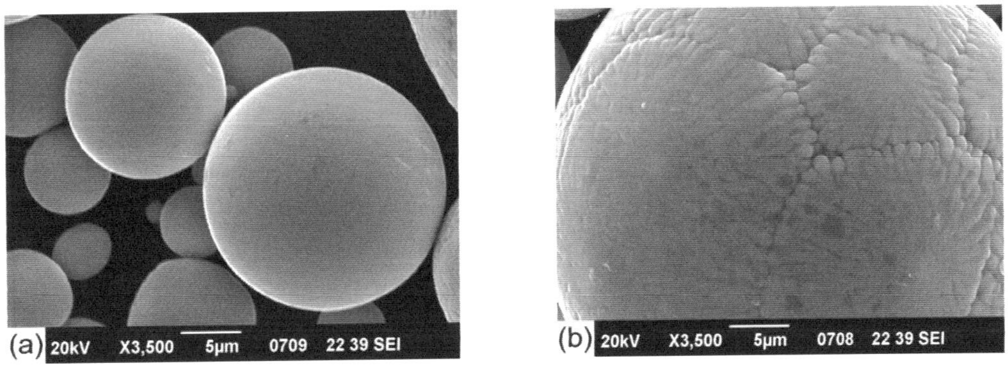

Figure 1. SEM micrographs in the secondary electron mode for powder particles of the studied alloy.

For particles of a satisfactory spherical shape, the elongation parameter should not exceed 1.2 – this condition is fulfilled for the particles of all investigated MPC fractions. The form factor for them is also at a fairly good level. Thus, it may be concluded that the shape of the powder particles is predominantly spherical and that the number of defective particles is approximately the same in different MPC fractions.

Table 1. Particle elongation and shape parameters for different MPC fractions.

Fraction, μm	Elongation, l_{max}/l_{min}	Form factor, $P/\sqrt{(S)}$
10–63	1.09 ± 0.01	3.89 ± 0.01
40–100	1.10 ± 0.01	3.96 ± 0.01
80–120	1.08 ± 0.01	3.94 ± 0.01

The results of technological properties measurements (flowability, apparent density, angle of repose) of the three investigated MPC fractions are presented in table 2.

Table 2. Technological properties for different MPC fractions.

Fraction, μm	Characteristics		
	Angle of repose, deg.	Apparent density, g/cm^3	Flowability, sec
10–63	43.7	1.960	-
40–100	34.9	2.124	40.0
80–120	31.3	2.164	38.3

The flowability of a powder directly depends on the powder particles shape and size, and the number of defects in the powder volume. According to table 2, the following relationship can be traced: with a decrease in the size of the powder fraction, the angle of repose increases, which affects flowability of the powder, reducing it. This is directly related to the particle size, since the number of defects in all the three investigated fractions is approximately the same. The lower is the angle of repose, the better is the powder flowability [15]. For free flow the flowability should be in the range of 30–40 sec, and the MPC fractions of 40–100 and 80–120 μm satisfy this condition. The angle of repose for the fraction of 10–63 μm is much higher – this powder has no flowability. It also has the lowest apparent density values.

Thus, the MPC fractions of 40–100 and 80–120 μm meet the flowability requirements for powders used in the synthesis of parts by selective electron-beam melting (SEBM) technology (no more than 40 sec per 50 g of powder through a calibrated Hall funnel). On the other hand, the MPC fraction of 10–63 μm could be used for the synthesis of parts by selective laser melting (SLM) technology.

The qualitative XRD analysis has revealed that the phase composition of the electrode is represented by the three phases corresponding to the equilibrium composition of the alloy in the as-cast state:

- γ-TiAl (PDF-2 # 01-081-9785);
- α_2-Ti_3Al (PDF-2 # 00-052-0859);
- β_0(B2) (PDF-2 # 00-044-1288).

The phase composition of the MPCs is represented by the three non-equilibrium phases:

- γ-TiAl (PDF-2 # 01-081-9785);
- α-Ti/α_2-Ti_3Al (PDF-2 # 00-044-1294, # 00-052-0859);
- β-Ti (PDF-2 # 00-044-1288).

The results of the phase analysis indicate that the process of atomization and subsequent cooling of the powder particles occurs at rates exceeding the critical cooling rates for the studied TiAl-based alloy, which leads to the formation of a quenched metastable state in the MPCs at room temperature. During the subsequent additive synthesis and heat treatment, the quenched state should transform into the equilibrium one.

The chemical composition analysis was carried out on samples taken from the initial electrode and MPCs. The comparative results are presented in table 3.

Table 3. Chemical compositions of electrode and MPC

Element	Content of elements, wt.%			
	In electrode		In MPC	In patent
	Upper part	Lower part		
Main alloying elements				
Ti	Bal.	Bal.	Bal.	Bal.
Al	30.92	31.07	30.80	30.00…35.00
V	2.63	2.65	2.55	1.0…3.0
Nb	2.47	2.45	2.32	1.5…4.0
Cr	2.46	2.49	2.57	2.0…3.0
Gd	0.37	0.40	0.39	0.2…0.6
Impurities				
C	0.0083	0.0065	0.0067	-
S	0.0043	0.0058	0.0010	-
O	0.034	0.027	0.043	-
N	0.011	0.0068	0.0019	-

4. Conclusions

The accomplished studies have shown that by gas atomization of the initial cast electrode, it is possible to obtain metal powder from a new Russian intermetallic β-solidifying TiAl-based alloy Ti-31.0Al-2.5V-2.5Nb-2.5Cr-0.4Gd (wt.%) of various fractional compositions (MPCs).

The chemical composition of the MPCs for the main alloying elements corresponds to that of the electrode; a 1.5-fold increase of the oxygen content in the MPCs is caused by natural oxidation of powder particles upon air environment due to developed specific surface.

It has been determined that the phase composition of the MPCs (γ+α(α_2)+β) differs from the equilibrium phase composition of the electrode (γ+α_2+β_0/B2) and corresponds to a rapidly quenched metastable state, which indicates high solidification rates in the atomization process, exceeding critical cooling rates of the studied gamma alloy.

High quality indicators have been achieved for the MPCs in terms of flowability and sphericity. Nevertheless, the powder flowability was found to be improved for the MPC fractions of 40–100 and 80–120 μm, making them applicable for additive synthesis of parts from the new TiAl-based alloy by SEBM method.

Acknowledgements
The research project was supported by the grant of the Russian Science Foundation No. 18-79-10249, https://rscf.ru/en/project/18-79-10249/.

The work was performed as a part of the implementation of the complex scientific direction 7.1 "Intermetallic Alloys Based on Titanium" ("Strategic Directions for the Development of Materials and Technologies of Their Processing for the Period until 2030") [16].

References

[1] Gibson I, Rosen D W and Stucker B 2015 *Additive Manufacturing Technologies* (New York: Springer Science and Business Media, LLC) 498 p
[2] Murr L E *et al.* 2012 *J. Mater. Sci. Tech.* **28** 1–14
[3] Kothari K, Radhakrishnan R and Wereley N M 2012 *Prog. Aero. Sci.* **55** 1–16
[4] Appel F, Paul J D H and Oehring M 2011 *Gamma Titanium Aluminide Alloys: Science and Technology* (Weinheim: Wiley-VCH Verlag & Co, KGaA) 745
[5] Wu X 2006 *Intermetallics* **14** 1114–22
[6] Wu X, Hu D and Loretto M H 2004 *J. Mater. Sci.* **39** 3935–40
[7] Kim Y-W and Kim S-L 2018 *J. Metals* **70** 553–60
[8] Wartbichler R, Clemens H and Mayer S 2019 *Adv. Eng. Mater.* **21** 1900800
[9] Mayer S, Erdely P, Fischer F D, Holec D, Kastenhuber M, Klein T and Clemens H 2017 *Adv. Eng. Mater.* **19** 1600735
[10] Clemens H and Mayer S 2013 *Adv. Eng. Mater.* **15** 191–215
[11] Wimler D, Lindemann J, Reith M, Kirchner A, Allen M, Vargas W G, Clemens H and Mayer S 2021 *Intermetallics* **131** 107109
[12] Panin P V, Nochovnaya N A, Lukina E A and Kochetkov A S 2019 *Inorg. Mater.: Appl. Res.* **10** 316–21
[13] Spitans S, Franz H and Baake E 2019 *Proc. of 11th PAMIR International Conf. – Fundamental and Applied MHD* 327–31
[14] Panin P V, Kochetkov A S, Zavodov A V and Lukina E A 2020 *Intermetallics* **121** 106781
[15] Neikov O, Naboychenko S and Dowson G 2009 *Handbook of Non-ferrous Metal Powders* (Elsevier: Technologies and Applications) 621 p
[16] Kablov E N 2015 Innovatsionnye razrabotki FGUP "VIAM" GNTs RF po realizatsii "Strategicheskikh napravleniy razvitiya materialov i tekhnologiy ikh pererabotki na period do 2030 goda" [Innovative developments of FSUE "VIAM" SSC of RF on realization of "Strategic Directions for the Development of Materials and Technologies of Their Processing for the Period until 2030"] *Aviacionnye materialy i tehnologii* **1(34)** 3–33 (in Russian)

Journal of Physics: Conference Series **2144** (2021) 012005 doi:10.1088/1742-6596/2144/1/012005

Rheological properties of powder blend for extrusion of ceramic-polymer filament used in 3D printing

A Smirnov, E V Kuznetsova, N N Babushkin, Y O Pristinskiy and N W Solis Pinargote

Laboratory of 3D Structural and Functional Engineering, Moscow State University of Technology "STANKIN", 127055, Moscow, Russia

E-mail: a.smirnov@stankin.ru

Abstract. The article presents the results of comparative studies of the rheological properties of the ceramic polymer blend of polylactide (PLA) filled with 50 %vol alumina to evaluate the possibility of obtaining extruded filament for 3D printing.

Fused deposition modeling (FDM) is one of the cheapest and available layer-by-layer techniques in the additive manufacturing domain [1]. Being printed out, the objects are ready for work right away without further processing. Although polymers are the main materials the technology works with, the latest trends include ceramics as well in the list of prospective materials [2–4]. Today, several scientific groups are conducting research on obtaining ceramic objects by FDM through developing ceramic-polymer filaments [2–6]. There are various methods for making filament for 3D printing, such as single-screw extrusion, twin-screw extrusion, chemical solvent method, etc. Single-screw extrusion is one of the cost-effective methods of obtaining raw materials, which consists of an inexpensive extruder and a winding unit. The raw materials prepared by extrusion must have acceptable rheological properties to avoid obstructions in the nozzle causing and prevent defects during printing.

The current work aims to study the rheological properties of the polymer compositions based on PLA filled with ceramic powder and to establish their optimal processing temperature for extrusion during FDM.

The study objects in this work are commercially available PLA powder sized d_{50}=35 µm and a ceramic-polymer blend of PLA and 50 %vol. alumina with particle size d_{50}=38 µm mixed in 1:1 ratio volume.

Rheology tests were performed on a rotational rheometer MARS (Thermo Fisher Scientific Inc., USA) with a "plane-plane" working unit geometry (diameter 20 mm, 1 mm offset). According to the recommended temperature range for the materials, the measurements were taken in the range 200-220°C for the PLA and 200-240°C the PLA-alumina blend.

Testing conditions were the following:

1. Frequency dependence of the dynamic modulus components at a deformation amplitude of 1% (the deformation frequency was varied from 0.1 Hz to 100 Hz).
2. Creep tests at a constant load of 10, 100 and 500 Pa for an unfilled PLA with subsequent removal of the load (recovery stage).
3. Dependence of viscosity on shear rate (the shear rate range was increased stepwise from 0.001 to 10 s^{-1} with holding at each stage for 30 s).

Content from this work may be used under the terms of the Creative Commons Attribution 3.0 licence. Any further distribution of this work must maintain attribution to the author(s) and the title of the work, journal citation and DOI.
Published under licence by IOP Publishing Ltd

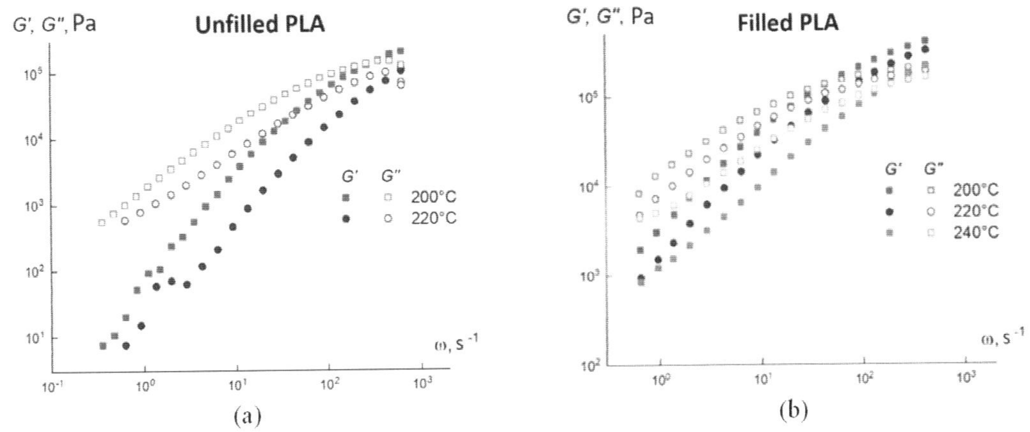

Figure 1. Frequency curves of the elastic modulus and loss modulus at the various temperatures for the unfilled PLA (a) and for the PLA filled with Al_2O_3 50 %vol. (b).

The study of the filled PLA showed that of rheological properties differs markedly from those of the unfilled PLA. Table 1 and figure 1b present viscoelastic behavior of the filled PLA composition in the form of frequency dependences. In contrast to the unfilled PLA (figure 1a), the frequency curves intersect at 60 s^{-1}, above which elastic deformations predominate.

Table 1. Viscoelastic behavior of the filled PLA.

Temperature, °C	**tg *G'***	**tg *G''***	$\omega_{crossover}$, s^{-1}	$G_{crossover}$, Pa
200	1,25	0,75	55	114,2*10^3
220	1,06	0,82	86	94,3*10^3
240	1,18	0,68	150	108,3*10^3

The crossover point at 200°C corresponds to modulus values 8.75×10^4 Pa, while for the unfilled PLA it was found to be $14{,}5\times10^4$ Pa. Above this value, the elastic component of the dynamic modulus prevails over the loss modulus, which indicates that the elasticity of the material becomes dominant.

Figure 2 shows graphs of the dependence of viscosity on shear stress for unfilled (a) and filled with 50 vol.% Al2O3 (b) polylactide at different temperatures.

As the temperature rises, the values of the modules are expected to decrease, and the crossover point shifts to the region of higher angular frequencies. Moreover, under stationary shear conditions (figure 2b), the filled PLA behaves like a Newtonian liquid with a viscosity of 1.5×10^4 Pa·s at 200 °C, which is approximately 15 times higher than the viscosity of the unfilled sample. Such an increase in viscosity is due to the presence of a filler and the adsorption of polymer macromolecules on it.

To explain the dependence of viscosity on shear stress for both the unfilled and filled PLA at different temperatures, studies were additionally carried out using differential scanning calorimetry and thermogravimetric analysis. Thermogravimetric analysis was performed on a TGA/DSC3 + machine (Mettler Toledo, Switzerland) with an HSS2 sensor. The tests were carried out in a heating mode at a constant rate of 10 K/min, in the range from 25 to 1000 °C, in an air atmosphere, with a flow rate of 50 ml/min.

The results (figure 3) showed that the filler shifts the destruction temperature to higher temperatures (358°C compared to 320°C for the unfilled PLA).

The study of thermal transitions between the unfilled and filled PLA was carried out on a DSC3 + device (Mettler Toledo, Switzerland), with a liquid nitrogen cooling system, HSS9 + sensor. The results showed that the filled sample noticeably differs in properties from the pure PLA: crystallinity is suppressed, and the glass transition temperature shifts to higher temperatures (T_{vit} = 101-108°C).

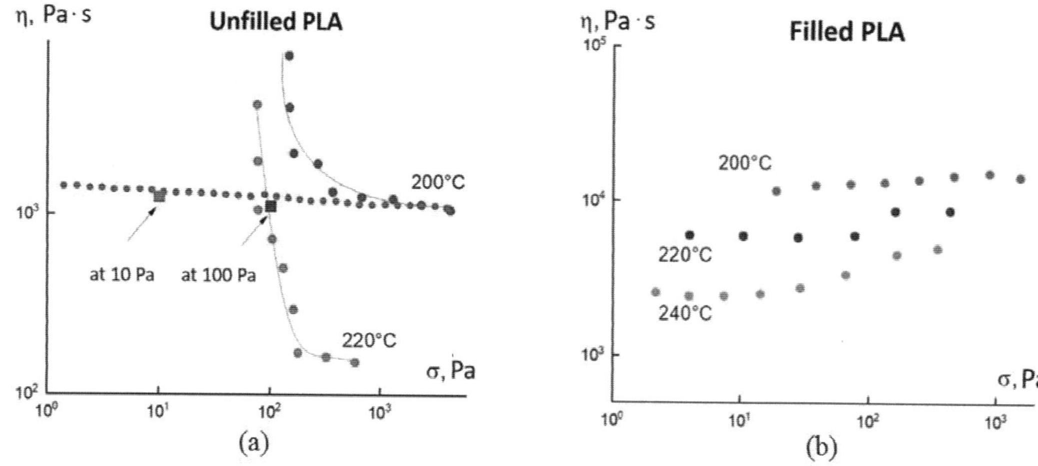

Figure 2. Shear stress versus viscosity plot for the unfilled (a) and filled (b) PLA at different temperatures.

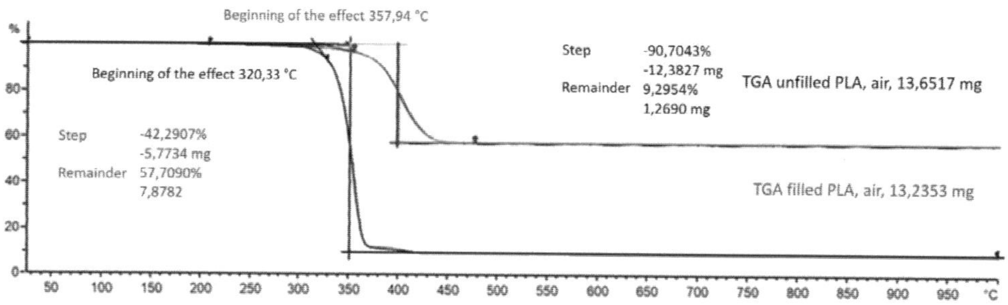

Figure 3. Thermogravimetric curve of the filled and unfilled PLA.

Overall, it was found that the rheological properties of the filled PLA are acceptable for the production of filament by extrusion. It behaves as a Newtonian viscoelastic liquid with viscosity much higher than that of unfilled PLA. This fact along with decreases angle coefficients of frequency dependencies of the elastic and loss modulus indicates the formation of a sufficiently strong structure in the composite. Further research will be aimed at increasing the content of the ceramic component in the mixture, obtaining filaments by extrusion and three-dimensional printing of products from them.

Acknowledgments
This work was supported by the Ministry of Science and Higher Education of the Russian Federation under project 0707-2020-0034. The study was carried out on the equipment of the Center of collective use of MSUT "STANKIN" (project 075-15-2021-695).

References

[1] Fawaz A 2016 *International Journal of Scientific & Engineering Research* **7(5)** 41–3
[2] Chen Z, Li Z, Li J, Liu C, Lao C, Fu Y, Liu C, Li Y, Wang P and He Y 2019 *Journal of the European Ceramic Society* **39(4)** 661–87
[3] Nötzel D, Eickhoff R and Hanemann T 2018 *Materials* **11(8)** 1463

[4] Veteška P, Hajdúchová Z, Feranc J, Tomanová K, Milde J, Kritikos M, Ľuboš B and Janek M 2021 *Applied Materials Today* **22** 100949
[5] Gorjan L, Galusca C, Marwah S, Sebastian T and Clemens F 2020 *Additive Manufacturing* **36** 101391
[6] Marsh A C, Zhang Y, Poli L, Hammer N, Roch A, Crimp M and Chatzistavrou X 2021 *Materials Science and Engineering* **118** 111516

Study of influence of micro-arc oxidation process duration on the characteristics of thermal control coatings for space applications

K A Anikin[1], A V Apelfeld[1,2] and I O Kondratskiy[1]

[1] Moscow State University of Technology "STANKIN", 127994, Moscow, Russia
[2] Moscow Aviation Institute (National Research University), 125993, Moscow, Russia

E-mail: airgear12@mail.ru

Abstract. Thermal control coatings were obtained on the AMg6 aluminum alloy with the aid of micro-arc oxidation (MAO) process. The dependences of the thickness, roughness, porosity and thermal control properties of the coating on MAO process duration were studied. The thermal control properties (solar absorbance α_s and emissivity ε) were investigated by ultraviolet–visible-near infrared spectrophotometer instrument and solar absorption reflectometer. The analysis of thickness, roughness and MAO process duration influence on the thermal control properties of the coating was carried out.

Thermal control coatings for space applications are ordinary of white or black color. Black coating prepare in the electrolyte based on Na_2SiO_3, Na_3PO_4, NaOH, EDTA-2Na and NH_4VO_3, Na_2WO_4, $KMnO_4$, $Cu(CH_3COO)_2$ [1–6] as well as graphene [7] as additives. Aluminum and its alloys are widely used in space instrumentation. However, these materials have some disadvantages, the effect of which can be reduced by additional surface treatment. One of the most promising methods for surface modification is micro-arc oxidation (MAO).

To study the effect of the duration of the technological process on the thermal control properties, MAO was carried out on samples of the AMg6 aluminum alloy in the anodic-cathodic mode with a frequency of 50 Hz at a total current density of 10 A/dm^2 and equality of the anodic and cathodic currents. The coatings were formed in accordance with the method described in the patent [8]. The MAO duration was varied from 30 to 135 min for white MAO-coating, and from 15 to 45 min for the black ones.

The thickness of the obtained coatings was determined using VT-201 eddy current thickness gauge. The thermal control properties (solar absorbance α_s and emissivity ε) were investigated by ultraviolet–visible-near infrared spectrophotometer instrument and solar absorption reflectometer. The dependences of white and black coatings thickness on duration of MAO process are shown in figure 1.

Figure 1 shows that the thickness of the optically black MAO-coating has an almost linear dependence on the processing time. At the same time, for white MAO-coating, a thickness difference is observed in the interval from 105 to 120 min.

The dependences of coatings roughness on the duration of MAO process are shown in figures 2 and 3. The roughness of white MAO-coating increases with an increase in the pores density on the coating surface and their diameter, and decreases with a decrease in the pores density, while maintaining their

Content from this work may be used under the terms of the Creative Commons Attribution 3.0 licence. Any further distribution of this work must maintain attribution to the author(s) and the title of the work, journal citation and DOI.
Published under licence by IOP Publishing Ltd

diameter. An increase in the size of globules formed by aluminum and silicon compounds also contributes to a decrease in roughness.

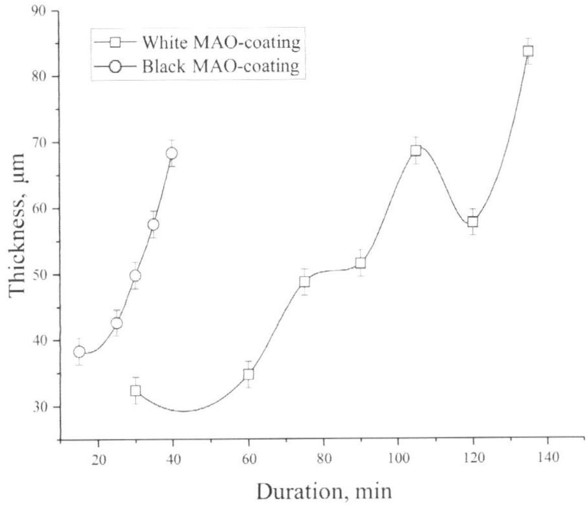

Figure 1. The dependence of coatings thickness on MAO treatment duration.

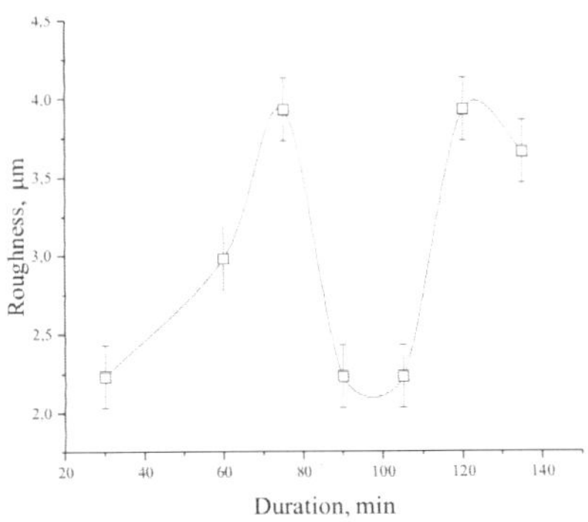

Figure 2. The dependence of roughness on MAO treatment duration for white coatings.

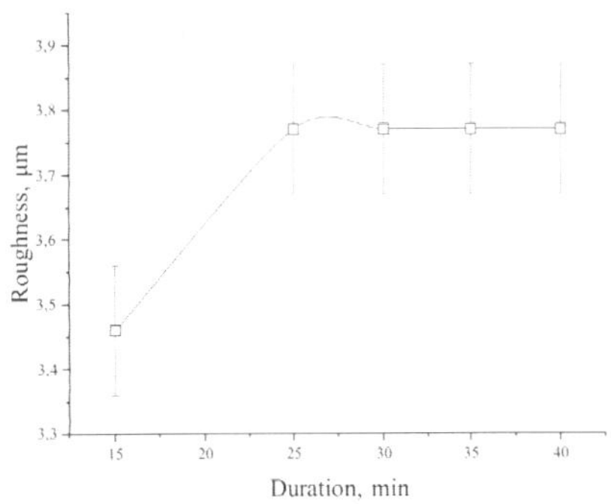

Figure 3. The dependence of roughness on MAO treatment duration for black coatings.

The roughness of black coating (figure 3) also increases at the initial stage, but becomes practically unchanged after 30 minutes of processing. This may be due to the fact that the surface does not have a pronounced globular morphology (figure 4), and the ratio of pore density and pore diameter is stabilized and remains unchanged with the continuation of processing.

Figures 5–7 show the dependence of thermal control properties (solar absorbance α_s and emissivity ε) of white coatings on its thickness and roughness.

The graphs presented in figures 5–7 show that with increasing thickness of the coating, solar absorbance α_s decreases and emissivity ε increases. At a thickness of more than 50 microns, the radiation from the alloy is completely absorbed by the coating. At the same time, as the roughness increases, α_s increases and ε decreases. This fact is explained by an increase in the total surface area, which allows more efficient heat dissipation, but also increases the absorption area of solar radiation.

At coating thicknesses less than 50 μm, the opposite picture is observed: with an increase in roughness, α_s decreases and ε increases. With a small thickness of the MAO-coating, an increase in the total area makes it possible to more efficiently absorb the radiation of the base material.

Figure 4. SEM image of black MAO-coating.

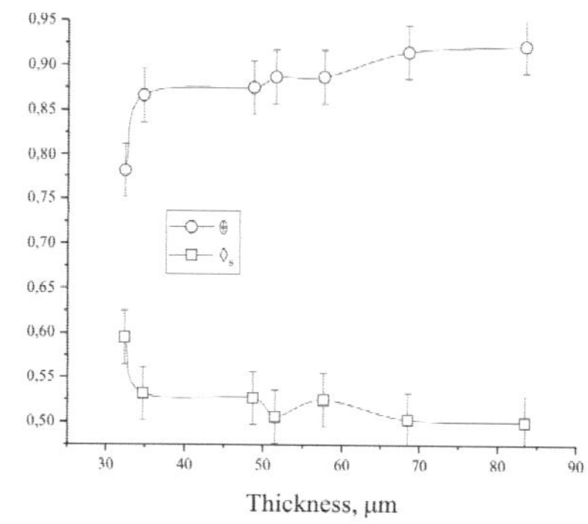

Figure 5. The dependence of thermal control properties on the thickness of white MAO-coatings.

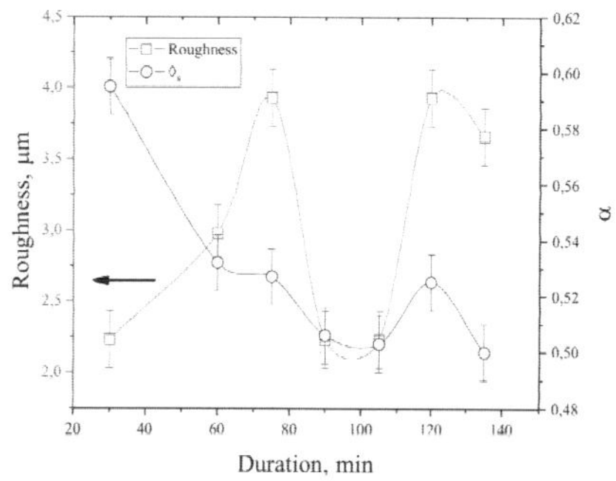

Figure 6. The dependences of roughness and α_s on MAO treatment duration for white coatings.

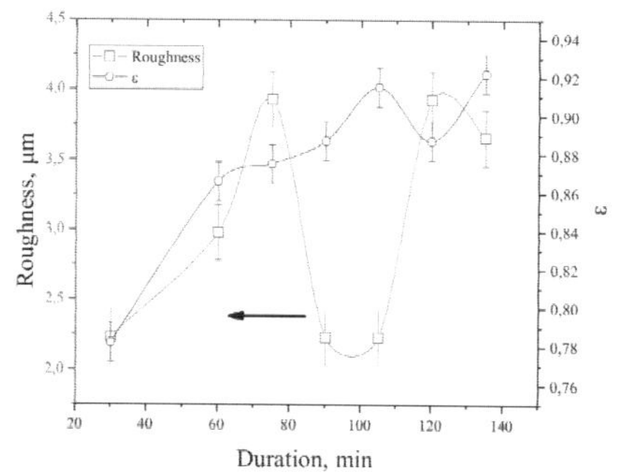

Figure 7. The dependences of roughness and ε on MAO treatment duration for white coatings.

Figures 8–10 show the dependence of the thermal control properties of black coating on the thickness and roughness.

Analyzing the changes in thermal control properties, thickness and roughness of black coating on the duration of the MAO-coating production process (figures 8–10), we can draw the following conclusions:

–at the initial stage, an increase in the thickness of the coating with an increase in the duration of treatment leads to a decrease in α_s and an increase in ε, and an increase in roughness corresponds to an increase and decrease in the coefficients of thermal control properties;

–after 25 min of treatment, an increase in the thickness of the coating leads to an increase in α_s and a decrease in ε (figures 9, 10), and the roughness does not affect these indicators, while it reaches Ra = 3.5 and does not increase further.

The change in thermal control properties is probably explained by the peculiarity of the formation of black MAO-coating. With MAO, vanadium oxide is incorporated into the structure of the upper layers of the coating, like other components of the electrolyte. But due to the relatively large particle size, it practically does not penetrate deep into the coating. At the initial stage, the coating layer containing vanadium compounds grows, which leads to a decrease in α_s and an increase in ε. With further processing, the growth rate of the "vanadium layer" slows down, and the release of components from the depth of the coating through the pores increases. As a result, there is a change in the thermal control properties of the coating towards the class of "solar reflectors" characteristic of white MAO-coating that does not contain vanadium.

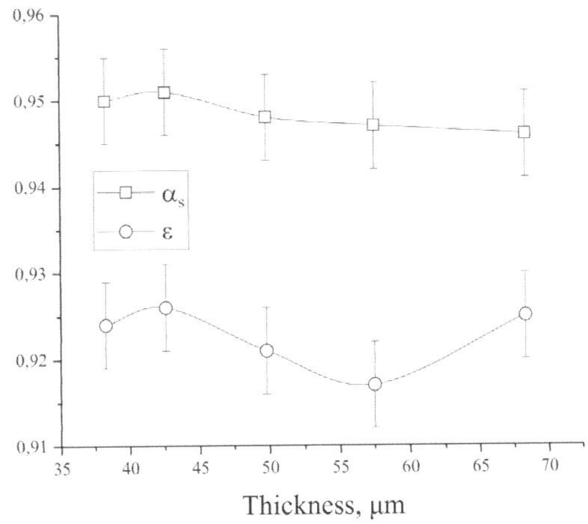

Figure 8. The dependence of thermal control properties on the thickness of black MAO-coatings.

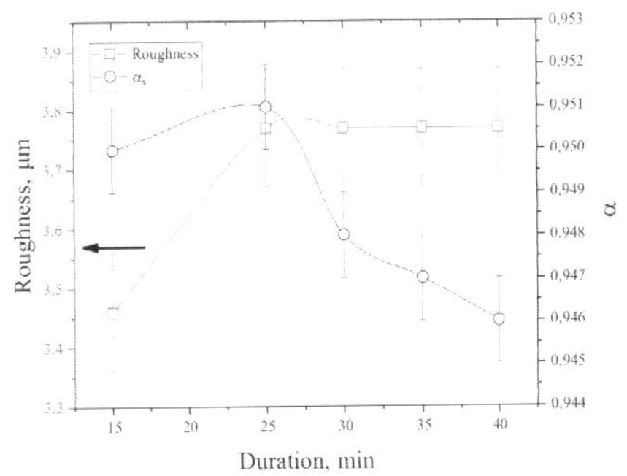

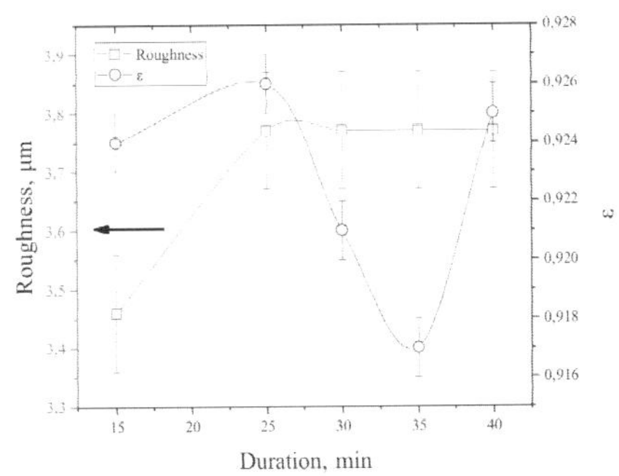

Figure 9. The dependences of roughness and α_s on MAO treatment duration for black coatings.

Figure 10. The dependences of roughness and ε on MAO treatment duration for black coatings.

The thickness of the coatings increases with the duration of MAO treatment.

The roughness of coatings is determined by the density of pores on the surface and their diameter and practically does not depend on the duration of MAO treatment.

As the thickness of white MAO-coating increases, solar absorbance α_s decreases and emissivity ε increases. Whereas with increasing roughness, α_s increases and ε decreases. However, at small coating thicknesses, the opposite picture is observed: with an increase in roughness, α_s decreases and ε increases.

At the initial stage of black coating forming, an increase in thickness leads to a decrease in α_s and an increase in ε, and an increase in roughness corresponds to an increase in the both coefficients of thermal control properties.

For black coatings after 25 min of MAO treatment, an increase in the coatings thickness leads to increase in solar absorbance α_s and decrease in emissivity ε.

Acknowledgments
The study was funded by the Russian Science Foundation (project number 21-79-30058).

The study was carried out on the equipment of the Centre of collective use «State Engineering Center" of MSUT "STANKIN" supported by the Ministry of Higher Education of the Russian Federation (project number 075-15-2021-695).

The authors are grateful to I V Suminov (Moscow State University of Technology "STANKIN") for instrumental assistance and B L Krit (Moscow Aviation Institute (National Research University)) for consultations on aerospace problems.

References

[1] Chunyan D, Hui Z, Zhiyong D and et al. 2019 *Mater. Lett.* **236** 723
[2] Tang H, Sun Q, Xin T, Yi C, Jiang Z and Wang F 2012 *Curr. Appl. Phys.* **12** 284
[3] Al Bosta M M, Ma K-J and Chien H-H 2013 *Infrared Phys. Technol.* **60** 323–34
[4] Wang Y M, Tian H, Shen X E, Wen L, Ouyang J H, Zhou Y and et al. 2013 *Ceram. Int.* **39** 2869
[5] Jin F-Y, Wang K, Zhu M, Shen L-R, Li J, Hong H-H and Chu P K 2009 *Mater. Chem. and Phys.* **114** 398
[6] Mohannad M S, Al Bosta M M and Ma K-J 2014 *Infrared Phys. Technol.* **67** 63
[7] Grigoriev S, Peretyagin P, Smirnov A, Solis W, Diaz L A, Fernandez A and Torrecillas R 2017 *J. Eur. Ceram. Soc.* **37** 2473
[8] Anikin K A, Borisov A M, Zheltukhin A V, Zhukov A A, Savushkina S V, Fedichkin I D, Chernik V N and Apelfeld A V 2018 *J. Surf. Invest.* **12(3)** 531

Effect of nanocoating morphology on the signal of X-ray Photoelectron Spectroscopy

V P Afanas'ev, L G Lobanova, D N Selyakov and M A Semenov-Shefov

National Research University «Moscow Power Engineering Institute», 111250, Moscow, Russia

E-mail: v.af@mail.ru

Abstract. The paper considers the application of the traditional X-ray photoelectron spectroscopy (XPS) methodology: the Overlayer Thickness Determination for the analysis of coating parameters. In particular situations considered in this work, it is energetically favorable for the atoms of the coating to form clusters, but not be evenly distributed on the surface of the substrate material. The change in the XPS signal is analyzed in situations when the coating is not a plane-parallel homogeneous layer, but an island (cluster) structure. The mathematical model of the XPS signal formation is considered for the case of the cluster covering in the form of parallelepipeds. Photoelectron path distributions (in the coating material) analysis indicated a strong dependence of the signal on the viewing angle. For the purpose of analysis, experimental spectra were obtained for several samples: gold depositions of various thicknesses on a silicon substrate. The spectra were measured for different viewing angles of photoelectrons and interpreted within the Straight Line Approximation (SLA). It is shown that proposed simplest model of an island coating allows to describe the effect of a decrease in the value of the effective average coating thickness, determined in plane-parallel geometry, with an increase in the viewing angle, observed in XPS experiments with angular resolution.

The paper presents the results of theoretical and experimental studies of coatings thicknesses ranges from units to fractions of angstrom, based on XPS Peak Shape Analysis. Direct use of traditional formulas obtained in the Strait Line Approximation [1] can lead to significant errors in determining the coating thicknesses.

Coating thicknesses determination based on X-ray Photoelectron Spectroscopy spectra analysis is a well-established procedure [1]. All methods for determining the thickness are based on formulas obtained in neglect of the processes of elastic scattering of a photoelectron. The influence of elastic processes on the flux density of registered photoelectrons is taken into account by introducing correction factors [2].

In a large number of cases, it is energetically favorable for coating atoms to combine into clusters [3], forming an island coating on the substrate surface. Most often, an island structure is observed if the fluence of the coating atoms is on the order of 10^{16} cm^{-2}, such an amount can form from one to several monolayers of a homogeneous coating.

We use the SLA method to calculate the intensity ratio of the characteristic peaks of the substrate and coating. For example, in the case of gold coating on silicon, we use peaks: Au4f7/2, Au4f5/2, Si2p3/2, Si2p1/2. The advantage of this choice lies in the proximity of the energies of the photoelectrons that form these peaks.

Content from this work may be used under the terms of the Creative Commons Attribution 3.0 licence. Any further distribution of this work must maintain attribution to the author(s) and the title of the work, journal citation and DOI.
Published under licence by IOP Publishing Ltd

Let us compare the ratios of Au and Si signals in the case of a uniform gold coating on silicon with a thickness d and islands occupying s $(0 < s \leq 1)$ on a unit area and having a height d/s, so that the coating volume is the same in both cases.

Consider the first case - a uniform flat cover. The XPS signal of the substrate (material 1), in accordance with [1], is determined by the formula:

$$Q_1(\mu_0,\mu,\varphi) = I_{h\nu} n_1 F_1(\mu,\mu_0,\varphi)\mu l_{in1} \exp\left(\frac{-d}{\mu l_{in1}}\right), \quad (1)$$

where $I_{h\nu} n_1 F_1(\mu_0,\mu,\varphi)\mu l_{in_1}$ – signal intensity of a semi-infinite layer of material 1; $\exp^{\left(\frac{-d}{\mu l_{in_1}}\right)}$ – attenuation of a signal in a layer of material 2; μ_0 - cosine on the incidence angle; μ - cosine of the reflection angle.

The intensity of the layer of material 2 is determined by the formula:

$$Q_2(d,\mu_0,\mu,\varphi) = I_{h\nu} n_2 F_2(\mu_0,\mu,\varphi)\mu l_{in2}\left[1 - \exp\left(\frac{-d}{\mu l_{in2}}\right)\right]. \quad (2)$$

In (1) and (2), the mean free path of electrons l_{in_1} in material 1 and l_{in_2} in material 2 practically coincide since the energies of the peaks of materials 1 and 2 are chosen so that the difference in energies will be much less than the energies of photoelectrons in the peaks of materials 1 and 2 themselves. Further $l_{in_1} = l_{in_2} = l_{in}$.

The thickness d of the deposited layer of material 2 will be determined based on the relative intensities of the peaks.

$$f(x) = \frac{Q_2(d,\mu_0,\mu,\varphi)}{Q_1(\mu_0,\mu,\varphi)} = \frac{F_2(\mu_0,\mu,\varphi) l_{in} n_2}{F_1(\mu_0,\mu,\varphi) l_{in} n_1}\left[\exp\left(\frac{-d}{\mu l_{in}}\right) - 1\right], \quad (3)$$

where d is the only unknown quantity.

Let us now consider the case of an island surface coverage. We analyze the unit area of silicon surface, in which $(1-s)$ silicon surface is free and - s part is covered with plane-parallel islands of height d/s, then formula (3), which determines the ratio of the intensities of the gold and silicon peaks will undergo noticeable changes.

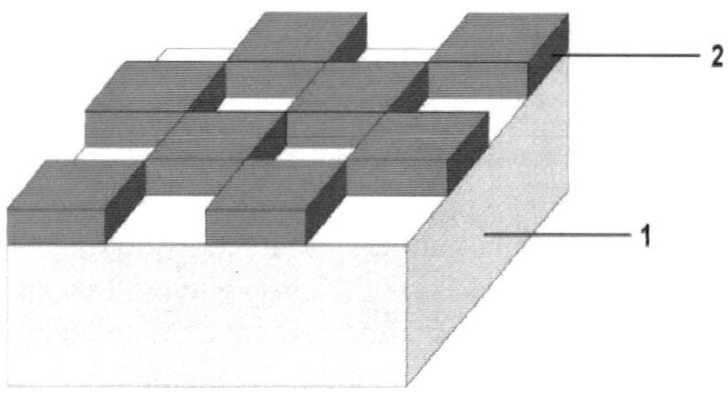

Figure 1. Island coating structure.

Substrate signal (material 1 - silicon):

$$Q_1(\mu_0,\mu,\varphi) = s I_{h\nu} n_1 F_1(\mu,\mu_0,\varphi)\mu l_{in} \exp^{\left(\frac{-d}{\mu s l_{in}}\right)} + (1-s) I_{h\nu} n_1 F_1(\mu,\mu_0,\varphi)\mu l_{in}. \quad (4)$$

Coating layer intensity (material 2 - gold):

$$Q_2(d,\mu_0,\mu,\varphi) = s I_{h\nu} n_2 F_2(\mu_0,\mu,\varphi)\mu l_{in}\left[1 - \exp\left(\frac{-d}{s\mu l_{in}}\right)\right]. \quad (5)$$

The relative intensity of the peaks for island coverage will lead us to the formula:

$$f1(x,s) = \frac{Q_2(d,\mu_0,\mu,\varphi)}{Q_1(\mu_0,\mu,\varphi)} = \frac{F_2(\mu_0,\mu,\varphi) n_2}{F_1(\mu_0,\mu,\varphi) n_1}\left[\frac{s*(1-\exp(\frac{-x}{s}))}{1-s+s*\exp(\frac{-x}{s})}\right], \quad (6)$$

where $x = d/\mu l_{in}$.

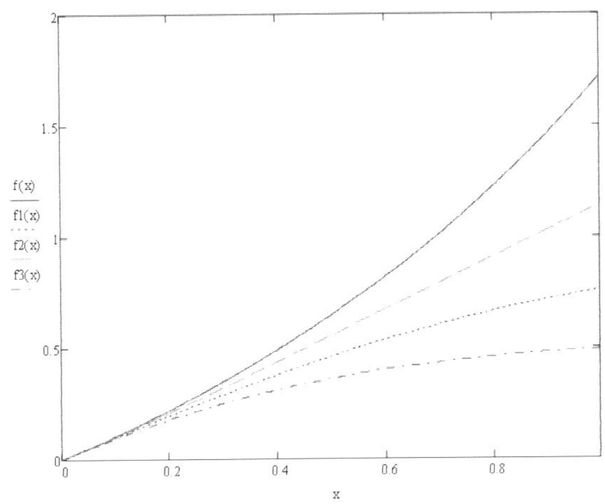

Figure 2. Dependence of the intensities ratio for the signals of the coating and the substrate on $x = 2d/\mu l_{in}$; $f(x)$ – corresponds to a uniform flat coating, $f1(x)$, $f2(x)$, $f3(x)$ – to an island structure, where $s = 0.7, 0.5, 0.35$.

Figure 2 is for the case of registration of photoelectrons along the normal to the target ($\mu = 1$). One should not use formulas (4) and (6) for other viewing angles. But formula (3) is valid at any angles of registration of photoelectrons. Let us show this in figure 3, where it can be seen that with an increase in the registration angle, the path of a photoelectron can pass not through one island of the coating, but through two, three or more. With an increase in the observation angle, the distribution of photoelectrons over the paths in the coating material in the case of an island coating ($s < 1$) and a uniformly distributed homogeneous layer ($s = 1$) approach each other.

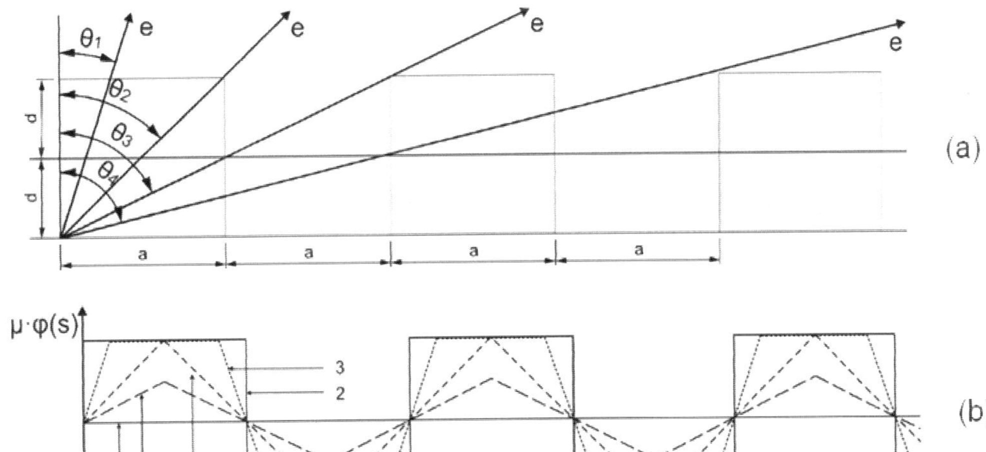

Figure 3. The path of the photoelectron depending on the angle of registration: (a) Photoelectrons trajectories emitted from the sample at different angles (SLA model). (b) Distribution of $\varphi(s)$ over the path lengths of photoelectrons in the coating as a function of the surface location of the photoelectron exit point: 1 - uniform coating with thickness d; 2,3,4,5 - exit from the island structure (formula 6), respectively, along the normal to the surface ($\theta = 0$) and at the angles θ_1, θ_2, θ_3, θ_4.

XPS measurements of submonolayer coatings of gold on silicon were carried out on two installations: on the SPECS installation in the AIC laboratory (Analytical Instrumentation Center) at the Vienna Technical University (TU Wien) with an energy analyzer capable of recording XPS spectra of photoelectrons emitted in the angular range from 27 to 75 degrees, and (B) on the KRATOS installation in the angle resolved mode - rotating the target to register XPS spectra at angles close to the normal and angles that are about 70° with the normal.

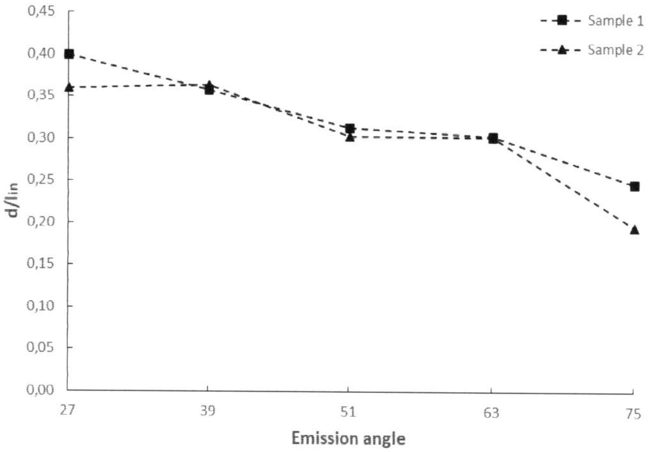

Figure 4. Gold layer coatings thickness values measured on the SPECS installation.

The experimental data processing was carried out on the basis of the methods presented in the works [4–6]. Calculations consistently showed a decrease in coating thicknesses values with an increase in the viewing angle relative to the normal. Similar measurements of the thicknesses of plane-parallel, uniform coatings led to the thickness values independent of the viewing angle.

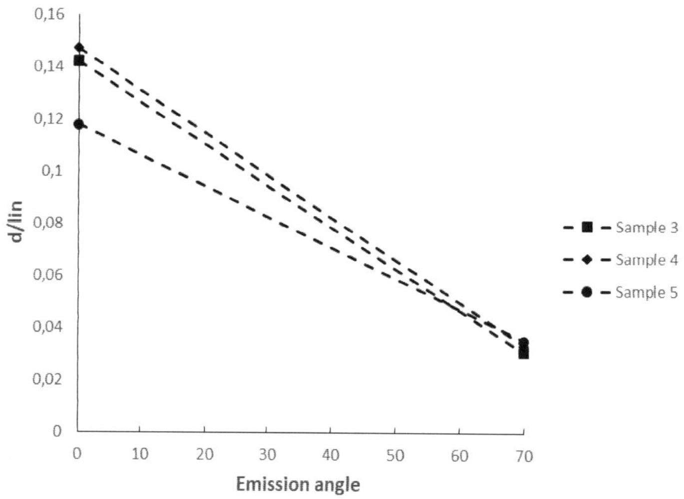

Figure 5. Gold layer coatings thickness values measured on the KRATOS installation.

The observed effect is described on the basis of a parallelogram island coverage model. This model made it possible to observe the change in the Path Length Distribution Function (PLDF) of the emitted photoelectrons depending on the viewing angle. PLDF determines the intensity ratio of the coating and the substrate XPS signals.

Results obtained indicate that a series of experiments with angular resolution should be carried out to determine the average effective coating thickness. The value of the critical viewing angle, after which the thickness value is stabilized, is determined by the density distribution of the islands *s*.

Acknowledgments
The research was performed in National Research University «MPEI» thanks to the financial support of ministry of science and higher education of RF, state assignment № FSWF – 2020 – 0023.

References

[1] Hoffman S 2013 *Auger and X-Ray Photoelectron Spectroscopy in Material Science* (Berlin: Springer Verlag)
[2] Afanas'ev V P, Efremenko D S and Kaplya P S 2018 *J. Surf. Invest.: X-Ray, Synchrotron Neutron Tech.* **12** 1182
[3] Smirnov B M 2017 *UFN* **187** 1329
[4] Jablonski A 2019 *Surf. Sci.* **688** 14–24
[5] Jablonski A and Zemek J 2009 *Surf. Interface Anal.* **41** 193–204
[6] Jablonski A 2012 J. *Phys. D: Appl. Phys.* **45** 14

Investigation of propagation dynamics of material deformations caused by laser pulse action

A F Banishev

Institute on Laser and Information Technologies – Branch of the Federal Scientific Research Centre "Crystallography and Photonics" of Russian Academy of Sciences, Shatura 140700, Moscow Region, Russia

E-mail: banishev@mail.ru

Abstract. The mechanoluminescent materials attract increasing attention of scientists due to their capability of visualizing the mechanical stresses and deformations experienced by them. The deformations of materials arising under the action of powerful laser pulses were studied. The composite mechanoluminescent materials based on the polymer and phosphor powder were used for visualization and registration of deformation evolution dynamics. The mechanoluminescent materials were deposited on the surface of the materials under study. It has been shown that the spatial distribution of glow intensity of the mechanoluminescent layer and the rate of its change make possible judging the value and rate of material deformation under laser pulses.

1. Introduction

The development of reliable and informative methods of measuring stresses and deformations in materials is of much interest for controlling the condition of various parts and structures exposed to the external forces (mechanical, electrical, or magnetic). To the known methods commonly employed in measurements of stresses and deformations belong strain gauges, photoelastic coatings, the digital image correlation method and the moiré interference method. Great interest has been recently expressed in the investigation of mechanoluminescent materials (mechano-phosphors) and in the development of sensors of deformations and stresses based on these materials. The interest has grown to studying mechano-piezo-phosphors and to the development of the new types of devices and mechanisms for mechano-piezo-photonics [1–15]. Mechano-phosphors are the materials transforming external mechanical actions to light emission. By now, a large number of them have been synthesized, which produce efficient mechanoluminescence in various spectral ranges [16, 17]. The mechanoluminescent coatings deposited on the surface of the component being studied are capable of visualizing the stresses and deformations in the complex-shape parts and over a large area, which cannot be done by other methods. Of special interest are the mechano-phosphors generating mechanoluminescence in the presence of elastic deformations over a wide range of loads. The mechanoluminescence intensity of the mechano-phosphors of this kind is known [18, 19] to be proportional to the amount of stresses and deformations of the surface, so they can be used in quantitative measurements of stresses and deformations in materials. The distribution of stresses and deformations in the material can be estimated by the distribution of glow intensity of the mechanoluminescent layer. In [20–22], crack formation and velocity in ceramics were studied with the use of a mechanoluminescent coating deposited on the sample surface. The formation and growth of

Content from this work may be used under the terms of the Creative Commons Attribution 3.0 licence. Any further distribution of this work must maintain attribution to the author(s) and the title of the work, journal citation and DOI.
Published under licence by IOP Publishing Ltd

cracks were initiated by means of an indenter. The application of mechanoluminescent coatings in the diagnostics of stresses and crack initiation and growth in materials under mechanical action proved to be highly informative. In [23] the mechanoluminescent coatings were employed for registration of deep microcracks initiated in the volume of material under cyclic loads and tension.

This work studies the deformations of metallic plates emerging on exposure to powerful laser pulses. The visualization of deformations was realized with the use of a composite mechanoluminescent material that was deposited on the plate surfaces. The dynamics of deformation was recorded by a high-speed video camera.

2. Materials and experimental procedure

To visualize the deformations, a composite mechanoluminescent material was prepared that was based on an adhesive transparent in the visible spectrum and a powder of the mechano-phosphor $SrAl_2O_4$:(Eu^{2+}, Dy^{3+}). A thin layer of the suspension was deposited on the surfaces of the metallic plates (stainless steel) of 100 to 200 μm thickness and ≈30 mm diameter. Hardening of the suspension yielded a layer of the composite material ≈150-200 μm thick. Figure 1 shows the scheme of the facility for investigation of the thermal deformations emerging under the action of powerful pulses of the YAG:Nd^{3+} laser (λ≈1.06μm, τ≈1.5 ms, W=1.5 J). The laser pulse was focused on the free surface of the metallic plate into a spot of d≈2-6 mm. This resulted in fast heating of the material in the area of radiation interaction with the material and caused its thermal deformation.

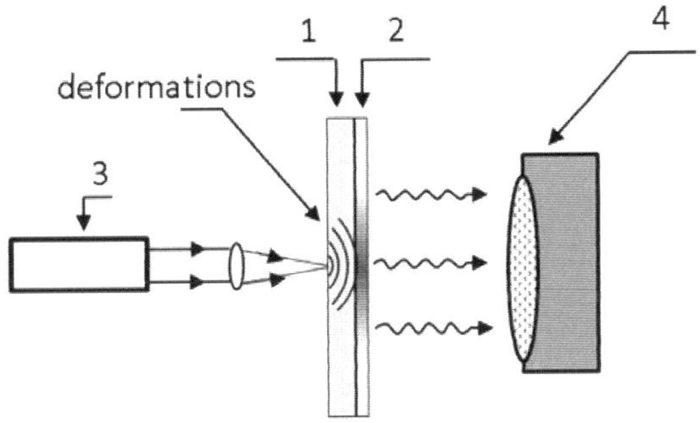

Figure 1. A scheme of the experimental facility. 1. Metallic plate. 2. Mechanoluminescent layer. 3. YAG:Nd^{3+} laser. 4. High-speed video camera.

3. Results and discussion

Estimate now the temperature T to which the backside of the plate will be heated at the end of laser pulse and the value of thermal stresses at r=1 mm, d=150 μm, W=1.5 J.

$$T(\tau_p) \approx \frac{AW}{\pi\left[r+(\chi\tau_p)^{1/2}\right]^2 d\rho c_m}, \tag{1}$$

where A≈0.35 is steel absorptance, χ=0.13 cm^2/s is thermal diffusivity, ρ=7.8 g/cm^3 is density, c_m=0.46 J/g•K is specific heat. Having substituted the numerical values, obtain T≈310 °C. To evaluate the thermal stresses arising by the end of the laser pulse in the irradiation area, use the expression:

$$\sigma(\tau_p) \approx \frac{\alpha G T(\tau_p)}{(1-\mu)}, \tag{2}$$

Substituting $\alpha=12.7*10^{-6}$ K^{-1} – thermal expansion coefficient of steel, $G=7.3*10^{10}$ N/m^2 – shear modulus, T≈310°C, μ=0.28 – Poisson ratio, we obtain $\sigma\approx 4*10^8$ N/m^2.

Figure 2 presents the views of the plate surface with the deposited mechanoluminescent layer, produced with the high-speed video camera (frame rate is 500 f/s).

The bright areas in the frames illustrate the mechanoluminescent layer glow caused by the deformations of the plate. Figure 2a shows the glow of the mechanoluminescent layer just after the end of the laser pulse. In the course of time, the luminous region rapidly increases in size, the rate of its growth far exceeding the velocity of temperature front propagation.

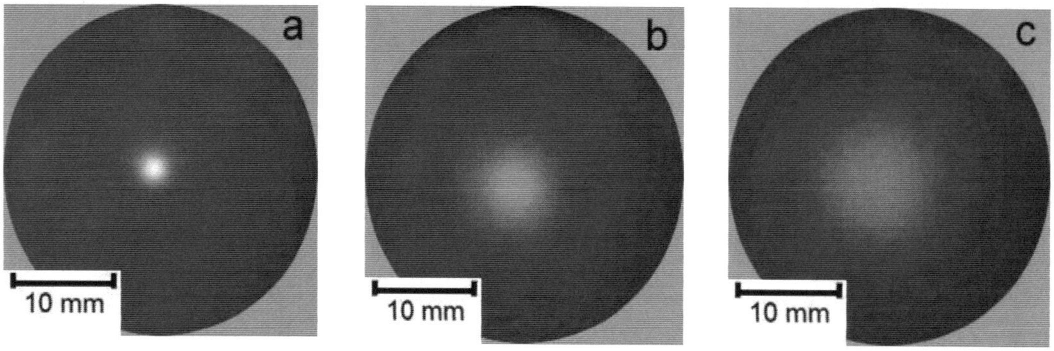

Figure 2. Visualization of deformations in the metallic plate by means of a mechanoluminescent layer: a) thermal deformations instantly after laser pulse action. Propagation of nonthermal deformations :b) in 15 ms, c) in 30 ms.

Figure 2b, c presents the images of the surface obtained in 15 ms and in 30 ms after laser pulse termination. This behavior demonstrates that the thermal deformations originated in the area of laser pulse action give birth to nonthermal deformations of the plate which propagate rapidly from this area. As the distance from the irradiation area increases, the glow intensity of the mechanoluminescent layer is lowered, thus pointing to a reduction in the amount of the plate deformations.

4. Conclusion

The investigations of material deformations arising under the action of powerful laser pulses were conducted. Visualization of the deformations was performed by using the composite mechanoluminescent materials deposited on the surface of the materials under study. It was shown that the glow distribution and intensity of the mechanoluminescent coatings give a rather comprehensive idea of the distribution rate and amount of material deformations emerging under laser pulses.

Acknowledgements

The reported study was funded by RFBR and ROSATOM, project number 20-21-00066 and partially supported by the RF Ministry of Science and Higher Education as a part of a State Task for the Federal Scientific Research Centre “Crystallography and Photonics” of Russian Academy of Sciences (in part of obtaining new nanomaterials and nanostructures for solving urgent problems of micro- and nanoelectronics and nanophotonics).

References

[1] Xiandi Wang, Dengfeng Peng, Bolong Huang, Caofeng Pan and Zhong Lin Wang 2019 *Nano Energy* **55** 389–400 DOI:org/10.1016/j.nanoen.2018.11.014

[2] Jun-Cheng Zhang, Xusheng Wang, Gerard Marriott and Chao-Nan Xu 2019 *Progress in Materials Science* **103** 678–742 DOI.org/10.1016/j.pmatsci.2019.02.001

[3] Ang Feng and Philippe F Smet 2018 *Materials* **11(4)** 484 DOI.org/10.3390/ma11040484

[4] Haitao Lv, Zhidong Pan and Yanmin Wang 2019 *J. Lumin* **209** 129–40 DOI:org/10.1016/j.jlumin.2019.01.026

[5] Chunfeng Wang, Dengfeng Peng and Caofeng Pan 2020 *Science Bulletin* **65(14)** 1147–9 DOI:org/10.1016/j.scib.2020.03.034

[6] Ali Imani Azad, Mohammad Reza Rahimi and Gun Jin Yun 2016 *Smart Mater. Struct.* **25** 095032 (14pp) DOI.org/10.1088/0964-1726/25/9/095032

[7] Srivatsava Krishnan, Hugo Van der Walt, Vijay Venkatesh and Vishnu Baba Sundaresa 2017 *Journal of Intelligent Material Systems and Structures* **28(17)** 2458–64 DOI: 10.1177/1045389X17689939
[8] Piyush Jha, Ayush Khare, Singh P K, Chandra V K and Sonwane V D 2018 *J. Lumin.* **195** 40–3 DOI.org/10.1016/j.jlumin.2017.10.063
[9] Gupta Karan Kumar, Kadam R M, Dhoble N S, Lochab S P, Vijay Singh and Dhoble S J 2016 *Journal of Alloys and Compounds* **688** 982–93 DOI:org/10.1016/j.jallcom.2016.07.114
[10] Puxian Xiong, Mingying Peng and Zhongmin Yang 2021 *iScience* **24** 101944 DOI:org/ 10.1016/j.isci.2020.101944
[11] Changjian Chen,Yixi Zhuang, Dong Tu, Xiandi Wang, Chaofeng Pan and Rong-Jun Xie 2020 *Nano Energy* **68** 104329 DOI:org/10.1016/j.nanoen.2019.104329
[12] Lejing Li, Lothar Wondraczek, Mingying Peng, Zhiwei Ma and Bo Zou 2020 *Nano Energy* **69** 104413 DOI.org/10.1016/j.nanoen.2019.104413
[13] Can Wang, Yun Yu, Yuhu Yuan, Chongyuan Ren, Qiuyan Liao, Jiaqiang Wang, Zhaofei Chai, Qianqian Li and Zhen Li 2020 *Matter* **2(1)** 181–93 DOI.org/10.1016/j.matt.2019.10.002
[14] Banishev A F and Banishev A A 2019 *Int. J. Modern Phys. B* 33(30) 1950366 DOI:10.1142/S0217979219503661
[15] Banishev A F and Banishev A A 2019 *JTF Letters* **45(10)** 3–6 DOI:10.21883/ PJTF.2019.10.47746.17737 (in Russian)
[16] Dong Tu, Xu C N, Yuki Fujio and Akihito Yoshida 2015 *Light: Science & Applications* **4** e356 DOI:10.1038/lsa.2015.129
[17] Kamimura S, Yamada H and Xu C N 2012 *Appl. Phys. Lett.* **101(9)** 091113 DOI:10.1063/ 1.4749807
[18] Chandra B P, Chandra V K and Jha Piyush 2015 *International Journal of Luminescence and Applications* **5(3)** 336–43
[19] Chandra B P, Chandra V K, Mahobia S K, Jha P, Tiwari R and Haldar B 2012 *Sensors and Actuators A: Physical* **173(1)** 9–16 DOI:10.1016/j.sna.2011.09.043
[20] Kim Ji Sik, Yong-Nam Kwon, Namsoo Shin and Kee-Sun Sohn 2007 *Applied Physics Letters* **90** 241916 DOI:10.1063/1.2748100
[21] Kim J S, Koh H J, Lee W D , Shin N, Kim J G, Lee K H and Sohn K S 2008 *Metals and Materials International* **14(2)** 165–9 DOI:10.3365/met.mat.2008.04.165
[22] Banishev A F 2021 *JTF Letters* **47** 11 DOI:10.21883/PJTF.2021.11.51005.18739 (in Russian)
[23] Fujio Yuki, Xu C N, Sakata Yoshitaro, Ueno Naohiro and Nao Terasaki 2020 *Journal of Alloys and Compounds* **832** 154900 DOI:org/10.1016/j.jallcom.2020.154900

Effect of TiN particles on mechanical and anticorrosive properties of oxide layers formed by PEO on silumin

E D Borgardt, A V Katsman and M M Krishtal

Research Division #4, Federal Government Budgetary Institution of Higher Education Togliatti State University, 445020, Togliatti, Russia

E-mail: euborg13@gmail.com

Abstract. The work focuses on the effect of sub-micron and micron dispersity titanium nitride additions to the electrolyte during plasma electrolytic oxidation (PEO) of silumin A14130 (11.5 wt% Si) on the structure, mechanical and anticorrosive properties of the oxide layers. It was found that the greatest positive effect of sub-micron TiN particles on the hardness and wear resistance of oxide layers is achieved at concentrations in the range of 1.0-2.5 g/l, and for micron TiN particles – at 0.5–1.0 g/l concentrations. However, for all concentrations of TiN particles, their insertion into the oxide layer has adverse effect on the anticorrosive properties of the oxide layer.

1. Introduction

Aluminum-silicon alloys (silumins) are among the most widespread cast non-ferrous alloys, and their use continues to grow in the last decade [1]. Al-Si-X alloys are used in mechanical engineering for the manufacture of internal combustion engine pistons, cylinder blocks and heads, various casings, wheel rims, pump parts, as well as kitchen accessories and medical equipment [1, 2]. Despite the widespread use, the insufficient surface properties of silumins (hardness and wear resistance, corrosion resistance in some environments) require additional surface treatment for their use in a number of applications [3–5]. A promising method of surface modification is the plasma electrolytic (microarc) oxidation technology (PEO or MAO), which allows to obtain ceramic wear- and corrosion-resistant oxide layers on non-ferrous alloys (Al, Mg, Ti, Zr, etc.) [6–8]. However, for silumins, the quality and properties of PEO-formed oxide layers are insufficient due to silicon being a silumin component, which negatively affects the breakdown and burning of microarc discharges [3, 9–11]. Therefore, there is a need in some metallurgical solutions to neutralize this negative effect.

Modification of PEO electrolyte with different additives of micron and nanoscale insoluble substances to control the properties of the forming oxide layers is a common practice [10, 12–15]. Titanium nitride (TiN) as a hard and heat–resistant material used for treatment of metal-cutting tool working parts is of particular interest as a modifier in PEO of silumins that also has a good history of modifying in PEO of magnesium alloys [16, 17].

The aim of this work is to elucidate the effect of adding TiN particles with different dispersion and concentration to the electrolyte on the rate of formation, structure and properties of oxide layers produced by PEO on silumins.

Content from this work may be used under the terms of the Creative Commons Attribution 3.0 licence. Any further distribution of this work must maintain attribution to the author(s) and the title of the work, journal citation and DOI.
Published under licence by IOP Publishing Ltd

2. Experimental

Oxide layers were obtained on 20×60×6 mm^3 specimens of silumin A14130 (11.5 wt% Si) after preliminary heat treatment as per the T6(8) [3] mode and grinding with SiC abrasive discs with grain size of 600-1500-2500-5000 grit. Before PEO, all specimens were washed in isopropyl alcohol and distilled water and dried. PEO was carried out in an electrolyte based on potassium hydroxide KOH (3 g/l), sodium pyrophosphate $Na_4P_2O_7 \times 10H_2O$ (7 g/l) and sodium metasilicate Na_2SiO_3 (10 g/l) aqueous solution for 120 minutes at a current density of 33 ± 1 A/dm^2. Different dispersity titanium nitride particles were used as additives: sub-micron (n-TiN, average particle dispersion, APS 400 nm) and micron-sized (m-TiN, APS ~ 2.5 μm) at concentrations from 0.5 to 2.5 g/l in 0.5 g/l increments. The process was carried out at a current frequency of 500 Hz, pulse duty factor of 65%, current density of 33±1 A/dm^2, current ratio in cathodic and anodic impulses of 1.2, and electrolyte temperature of 290 ± 2K.

The thickness, morphology and structural features of the obtained specimens were investigated using a scanning electron microscope (Jeol JCM-6000 Neoscope) on transverse sections. They were also used to determine microhardness per Vickers HV0.1 with a Shimadzu DUH-211S microhardness meter (at least 10 times). The elemental composition of oxide layers was studied by energy dispersive X-ray fluorescence analysis using a Shimadzu EDX-8000 spectrometer. Bragg-Brentano X-ray diffraction analysis of the specimens was performed in the angle range of 15-80°, speed of 0.25°/min^{-1}, 0.02° step size on a Shimadzu Maxima XRD-7000 diffractometer. Phase identification was done using the Shimadzu PDF2 database. Oxide layer quantitative phase composition was determined by the Rietveld method in the JANA 2006 software. Wear resistance was evaluated by the Nanovea TRB 50N tribometer according to ASTM G133-99 Type A recommendations (Ø 6.35 mm steel ball with hardness HRC 62...63 used as counterbody) similarly to paper, oxide layer mass wear W, g/(N*m) was measured. Corrosion resistance of the specimens was evaluated by the methods of potentiodynamic polarization and impedance spectroscopy in 3.5 wt% NaCl solution at room temperature using Elins P-45X potentiostat with an electrochemical impedance measurement module FRA-24M in a three-electrode Ag/AgCl electrochemical cell with a KCl saturated reference electrode and a platinum opposing electrode. Prior to taking measurements, the specimens were held in a corrosive environment for at least 60 minutes for the open circuit potential (OCP) stabilization. At least 3 tests were performed on each of the specimens.

3. Results

All oxide layers, except the case of the 0.5 g/l n-TiN addition (Figure 1), are characterized by good continuity and low porosity, lack of transverse or main cracks, and the absence of the oxide layer delamination from the substrate. Addition of n-TiN, even at the concentration of 0.5 g/l, resulted in an increase in the average thickness of the oxide layer from ~97 μm (base layer) to ~105 μm. With increasing concentration of n-TiN, the average thickness of the oxide layer reaches a maximum at 1.0 g/l (~ 110-120 μm) and then remains at the level of 105 ± 5 μm. Addition of larger particles (m-TiN) to the electrolyte leads to a smaller increase in the thickness of the oxide layer with increasing concentration: from ~98 μm at 0.5 g/l to 103±5 μm at 1.0–2.5 g/l. It should be emphasized that adding TiN particles of any dispersity to the electrolyte significantly increases the process productivity: a thickness of ~ 100 μm is achieved already after 60-80 minutes of the process versus 100-120 minutes for PEO in the base electrolyte.

The major phases of oxide layers are corundum α-Al_2O_3 (from 34 to 65 wt%), gamma-alumina γ-Al_2O_3 (26.5-65 wt%), and mullite (0.2-5.5 wt%). When TiN particles are introduced into the electrolyte, the TiN cubic face-centered phase is identified in the oxide layer, which corresponds to the initial TiN powder. The TiN content in the oxide layers ranges from 1.6 to 3.8 wt%, and the maximum TiN content is reached at a concentration of 1.5 g/l of particles in the electrolyte. A further increase of TiN concentration in the electrolyte leads to a certain decrease in the TiN amount in the oxide layer (up to 2.5-2.7 wt%). Introduction of TiN particles into the electrolyte leads to decreasing the oxide layer degree of crystallinity from 82 vol% to 72-78 vol.%, and the lowest degree of crystallinity is observed at the maximum concentration of TiN particles in the electrolyte.

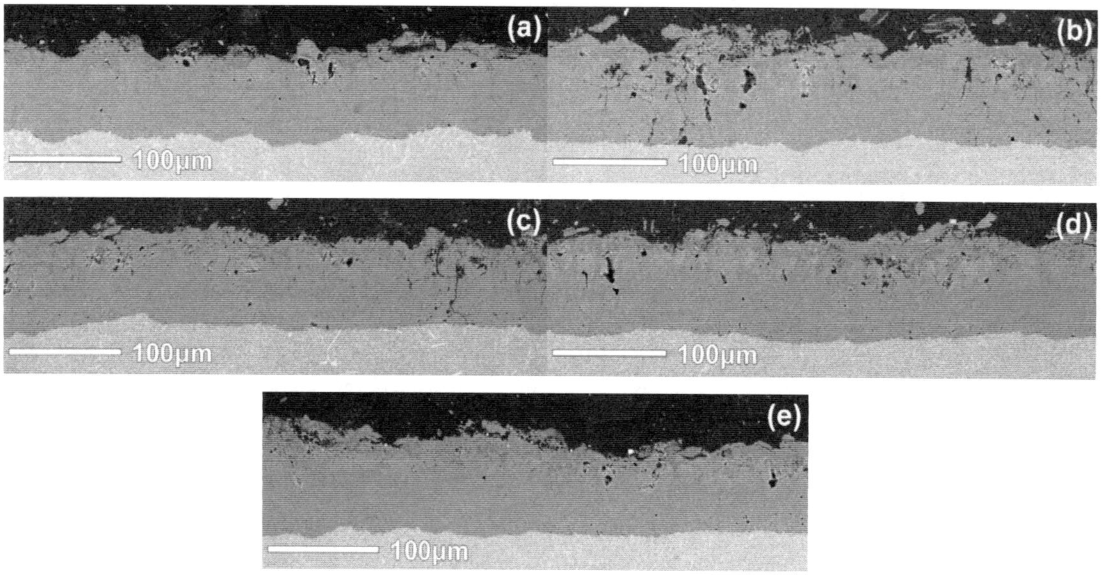

Figure 1. Cross-section of oxide layers (x200) with no addition (**a**) and with addition of TiN particle to the electrolyte: 0.5 g/l ((**b**) – n-TiN, (**c**) – m-TiN) and 2.5 g/l ((**d**) – n-TiN, (**e**) – m-TiN).

Addition of n-TiN and m-TiN particles to the electrolyte influenced the hardness (HV, MPa) and mass wear (W, ng/N·m) of oxide layers (figure 2). At n-TiN concentration of 1.5–2 g/l in electrolyte, there is an increase in average hardness up to ~910 MPa (by 35%), and a decrease in wear to 75 ng/N·m (by 67%) relative to the base specimen (650 MPa and 234 ng/N·m, respectively). Micron sized m-TiN particles demonstrated the effect of reducing wear and increasing hardness of the layers at a concentration of 1 g/l. Hardness increased to 969 MPa (by 39%), wear decreased to 125 ng/N·m (by 46.5%). It should be noted that the minimum wear was observed in the range of 0.5-1.0 g/l for both TiN powders added to the electrolyte. An increase in the concentration of m-TiN particles from 1.0 to 2.5 g/l led to a smoother decrease in hardness and an increase in wear values than in the sub-micron additive case. The lack of clear correlation of wear resistance and oxide layer hardness is worth noting.

Silumin treatment by PEO method triggers a shift of OCP to the region of more negative values (by about 0.5-0.6 V as compared to the initial alloy) (figure 3a). The addition of TiN particles to the electrolyte brings about a slight potential shift towards positive side (figure 3a). Despite the shift of E_{corr} to the region of more negative values, application of PEO allowed to reduce the corrosion current density i_{corr} as compared to the non-treated specimen by more than 3 times (from $5.1 \cdot 10^{-7}$ to $1.63 \cdot 10^{-7}$ A/cm^2) (figure 3b, table 1). The addition of titanium nitride powders into the electrolyte during PEO did not result in significant changes in the corrosion current densities (figure 3b, table 1). Protective effect of the obtained oxide layers is achieved by further anodic polarization (by 1V or more), under which all specimens are passivated and no alloy corrosion dissolution occurs. Polarization resistance of the specimens produced by PEO using base electrolyte increases by an order of magnitude versus the initial alloy. The effect is slightly worse for oxide layers obtained in an electrolyte with TiN particle additives: R_p increases by 2-3 times with addition of micron titanium nitride particles (m-TiN), and by 5 times with addition of titanium nitride sub-micron particles (n-TiN) versus non-treated alloy (table 1).

According to impedance spectroscopy data (Nyquist and Bode curves (Figure 4, a, b)), a rather high resistance of the natural oxide (barrier) film on silumin A14130 can be noted: impedance modulus $|Z|_{f=0.01\ Hz}$ reaches ~76 kΩ·cm^2. PEO in base electrolyte raises the value $|Z|_{f=0.01\ Hz}$ of the oxide layer by ~ 2 times – to 150-180 kΩ·cm^2. TiN particles added to the electrolyte during PEO reduced $|Z|_{f=0.01\ Hz}$ of the oxide layer to the bare alloy value, for any of the particle dimensions used. These findings are similar to the results obtained for the magnesium alloy [16].

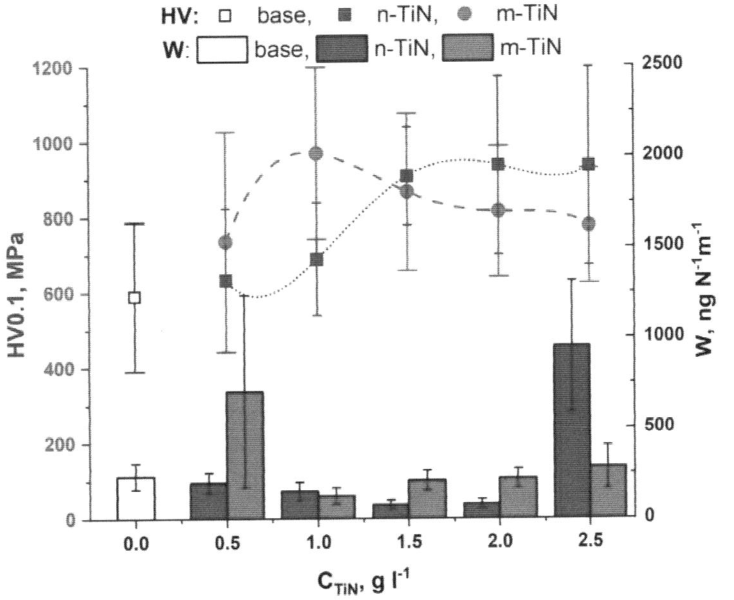

Figure 2. Hardness and wear resistance of oxide layers obtained by PEO in the base electrolyte and with additions of sub-micron (n-TiN) and micron sized (m-TiN) titanium nitride particles at different concentrations in the electrolyte.

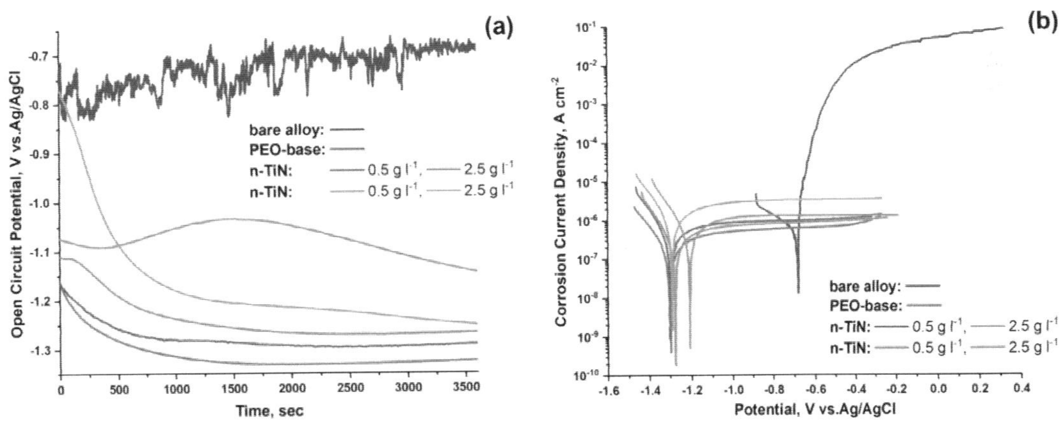

Figure 3. OCP evolution during 1h exposure in a 3.5% NaCl solution (a) and Tafel curves (b) of the substrate (bare alloy), with a base oxide layer (PEO-base) and with oxide layers with addition of sub-micron (n-TiN) and micron-sized (m-TiN) titanium nitride particles (0.5 and 2.5 g/l).

Apparently, despite the lack of obvious through macrodefects of the structure, the obtained oxide layers have a large number of through micropore channels down to the oxide layer/aluminum substrate interface. The main protective effect is apparently provided by a thin barrier layer at this interface which is not altered by the TiN particles. It can be assumed that with prolonged holding of specimens in a corrosive environment, micropore channels are filled with corrosion products, and the resistance to charge transfer grows. It should be noted that introduction of TiN particles to the electrolyte entails an increase in the inhomogeneity of anticorrosive properties of oxide layers, which is probably associated with the inhomogeneous distribution of particles in the structure of oxide layers, and is the subject of further research.

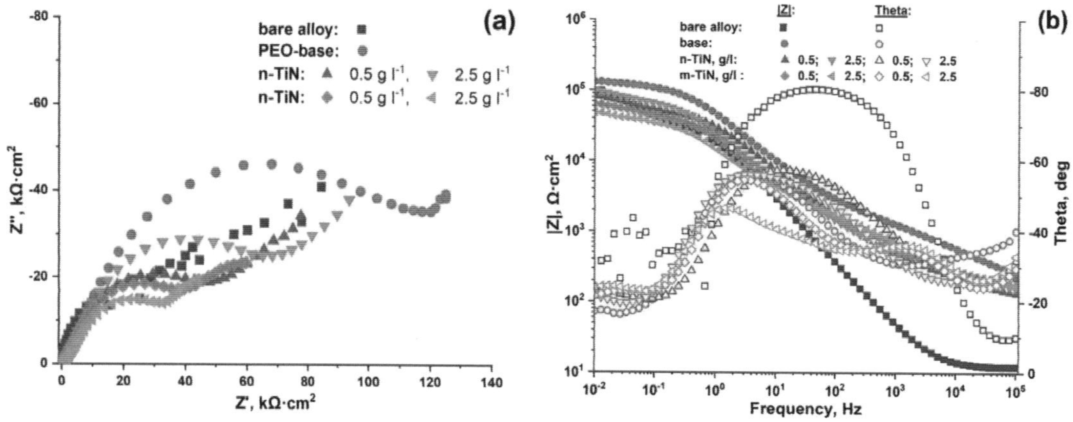

Figure 4. Nyquist (a) and Bode (b) plots for the alloy samples in the initial state (bare), after PEO in the base electrolyte and in the electrolyte with the additives of sub-micron (n-TiN) and micron sized (m-TiN) titanium nitride particles (0.5 and 2.5 g l^{-1}) in a 3.5% NaCl solution.

Table 1. Results of PDP and EIS tests of the bare Al-Si alloy, after PEO in the base electrolyte and in the electrolyte with the additives of sub-micron (n-TiN) and micron (m-TiN) titanium nitride particles (0.5 and 2.5 g l^{-1}) in a 3.5% NaCl solution at 22±2 °C[a].

Parameter / Electrolyte		OCP, V	i_{corr}, μA·cm^{-2}	E_{corr}, V	R_p, kΩ·cm^2	$\|Z\|_{f=0.01\ Hz}$, kΩ·cm^2
Bare alloy		-0.72 ± 0.02	0.51 ± 0.06	-0.68 ± 0.04	21 ± 11	76 ± 11
PEO-base		-1.28 ± 0.05	0.16 ± 0.03	-1.32 ± 0.03	224 ± 52	164 ± 29
n-TiN	**0.5**	-1.28 ± 0.03	0.39 ± 0.04	-1.28 ± 0.03	112 ± 59	63 ± 33
	2.5	-1.17 ± 0.08	0.66 ± 0.07	-1.19 ± 0.11	110 ± 70	67 ± 43
m-TiN	**0.5**	-1.18 ± 0.04	0.73 ± 0.05	-1.23 ± 0.02	61 ± 34	41 ± 20
	2.5	-1.26 ± 0.01	0.78 ± 0.02	-1.31 ± 0.01	48 ± 5	38 ± 9

[a] ±σ is given as an error

4. Conclusion

The findings of this work disclosed that the introduction of sub-micron or micron-sized TiN particles into the electrolyte triggers their inert insertion into the oxide layer during PEO. At the same time, these additions provide an increase in the rate of formation of oxide layer by 1.5-2 times, an increase in the hardness and wear resistance values up to 1.5 and 3 times, respectively, depending on the dispersity and concentration of TiN particles in the electrolyte. However, insertion of TiN particles into the oxide layer reduces its anticorrosive properties: there is a growth in the corrosion current density by up to ~5 times, and a drop in the impedance modulus $|Z|_{f=0.01\ Hz}$ by up to ~4 times compared to the oxide layer obtained in the electrolyte without addition of TiN particles. Thus, along with a significant increase in processing productivity and an improvement in the mechanical characteristics of the PEO formed oxide layers on silumin A14130 (11.5 wt% Si), a deterioration in the corrosion resistance is observed, which requires additional technological measures to eliminate this negative effect, and is the subject of further research.

Acknowledgements
The work was financed by a grant from the Russian Science Foundation (Project No. 21-19-00656). The authors are grateful to Mr. Anton Polunin (Togliatti State University) for assistance in performing X-ray structural analysis, and to Alisa Cheretaeva (Togliatti State University) for assistance in electrochemical studies.

References
[1] Michael V Glazoff, Khvan A V, Vadim S Zolotorevsky, Nikolai A Belov and Alan T Dinsdale 2019 *Casting Aluminum Alloys* (Elsevier)
[2] Lumley R N 2018 *Fundamentals of Aluminium Metallurgy* (Elsevier)
[3] Krishtal M M 2004 *Met. Sci. Heat Treat.* **46** 377–84
[4] Li K, Li W, Zhang G, Zhu W, Zheng F, Zhang D and Wang M 2019 *J. Alloys Compd.* **790** 650–6
[5] He J, Cai Q Z, Luo H H, Yu L and Wei B K 2009 *J. Alloys Compd.* **471** 395–9
[6] Yerokhin A L, Nie X, Leyland A, Matthews A and Dowey S J 1999 *Surf. Coatings Technol.* **122** 73–93
[7] Sikdar S, Menezes P V, Maccione R, Jacob T and Menezes P L 2021 *Nanomaterials* **11** 1375
[8] Simchen F, Sieber M, Kopp A and Lampke T 2020 *Coatings* **10** 628
[9] Mohedano M, Matykina E, Arrabal R, Mingo B and Pardo A 2015 *Appl. Surf. Sci.* **346** 57–67
[10] Polunin A V, Cheretaeva A O, Borgardt E D, Rastegaev I A, Krishtal M M, Katsman A V and Yasnikov I S 2021 *Surf. Coatings Technol.* **423** 127603
[11] Rogov A B, Lyu H, Matthews A and Yerokhin A 2020 *Surf. Coatings Technol.* **399** 126116
[12] Lu X, Mohedano M, Blawert C, Matykina E, Arrabal R, Kainer K U and Zheludkevich M L 2016 *Surf. Coatings Technol.* **307** 1165–82
[13] Polunin A V, Borgardt E D, Shafeev M R and Krishtal M M 2019 *Journal of Physics: Conference Series* **1396** (Institute of Physics Publishing) 012032
[14] Arrabal R, Matykina E, Skeldon P and Thompson G E 2008 *J. Mater. Sci.* **43** 1532–8
[15] Yao Z, Jiang Y, Jiang Z, Wang F and Wu Z 2008 *J. Mater. Process. Technol.* **205** 303–7
[16] Mashtalyar D V, Gnedenkov S V, Sinebryukhov S L, Imshinetskiy I M and Puz A V 2017 *J. Mater. Sci. Technol.* **33** 461–8
[17] Mashtalyar D V, Sinebryukhov S L, Imshinetskiy I M, Gnedenkov A S, Nadaraia K V, Ustinov A Y and Gnedenkov S V 2020 *Appl. Surf. Sci.* **503** 144062

Mechanical properties and corrosion resistance of Mg-RE cast alloys and their plasma electrolytic oxidation

S V Zasypkin, A O Cheretaeva, M R Shafeev, D L Merson and M M Krishtal

Research Division #4, Federal Government Budgetary Institution of Higher Education Togliatti State University, 445020, Togliatti, Russia

E-mail: a.cheretaeva@tltsu.ru

Abstract. The effect of heat treatment on the mechanical properties (hardness, plasticity, yield and tensile strength) and corrosion resistance of several cast magnesium alloys with additions of rare earth metals (Y, Nd and Gd), and their surface modification by plasma electrolytic oxidation (PEO) were investigated. It was found that the heat treatment of the alloys results in formation of $Mg_{12}YZn$, $Mg_3Zn_3Y_2$ and $Mg_{24}Y_5$ based LPSO-phases and causes an increase in hardness and tensile strength by 5-7 and 20-25%, respectively, but at the same time, corrosion resistance of the alloys drops by 10-20 times. PEO of the alloys after heat treatment reduced the corrosion currents by 1-3 orders of magnitude without changing the corrosion potential.

1. Introduction

Magnesium alloys are the lightest metal materials that are most in demand for structures requiring weight reduction, mainly in transport systems: aerospace, automotive, high-speed railway equipment. Accordingly, progressively higher requirements are imposed on such alloys every year, both in terms of strength characteristics, and technological and functional properties [1].

There are various ways to improve the properties of magnesium alloys, many of which (such as modern methods of intensive plastic deformation) are not applicable to cast alloys. There are three main ways to improve their properties: (1) modification of chemical composition (alloying); (2) optimization of casting technology and heat treatment; (3) surface modification (surface treatment, coating, etc.). In this regard, Mg alloys containing a strengthening long period stacking ordered phase (LPSO) that prevents the movement of dislocations and reduces the anisotropy of alloy properties, are considered the most prospective. However, the LPSO phase can significantly increase the corrosion rate of Mg alloys [2], therefore additional surface treatment is necessary to create an anticorrosive layer on the surface of such alloys. This problem can be successfully solved by plasma-electrolytic (microarc) oxidation (PEO or MAO) of Mg alloys [3–5].

2. Experimental

The heat treatment of alloys with rare earth (RE) elements, Mg-2.2Nd-0.6Zr-0.4Zn (ML10), Mg-6.8RE-2.5Zn-0.6Zr and Mg-6.8RE-2.5Zn-0.6Zr systems manufactured by SOMZ LLC (Solikamsk, Russia) was followed by MAO treatment, in the aim to improve their mechanical characteristics and corrosion resistance. For this purpose, standard specimens were manufactured (in accordance with the requirements of GOST 1497-84. "Metals. Tensile testing methods" that were tested on a universal

Content from this work may be used under the terms of the Creative Commons Attribution 3.0 licence. Any further distribution of this work must maintain attribution to the author(s) and the title of the work, journal citation and DOI.
Published under licence by IOP Publishing Ltd

tensile strength testing machine N50KT in the initial (cast) state and after heat treatment as per the T6 mode (homogenizing annealing at 540°C for 12 hours, quenching in air and complete artificial aging at 200°C for 12 hours). The Brinell hardness of the obtained specimens was also determined using Beijing Time Group HB-3000B hardness tester.

MAO of the alloys after heat treatment was performed in an alkaline-phosphate-aluminate-silicate-fluoride electrolyte (1 g/l KOH, 5 g/l $Na_2HPO_4 \times 12H_2O$, 5 g/l NaF, 7 g/l $Na_2SiO_3 \times 5H_2O$ and 7 g/l $NaAlO_2$) at a current frequency of 500 Hz, current density 11.5±0.5 A/dm^2, duty cycle factor 50%, and current ratio in the cathodic and anodic impulses 0.5 ± 0.05.

Confocal laser scanning microscopy, X-ray diffraction and X-ray fluorescence energy dispersion analysis were used to study the produced specimens of oxide layers on tested magnesium alloys. Confocal laser scanning microscope Olympus LEXT OLS 400 and scanning electron microscope Jeol Neoscope JCM-6000, X-ray diffractometer Shimadzu Maxima XRD-7000S and energy dispersive spectrometer Shimadzu EDX-8000 were used. The specimen corrosion resistance was determined by the method of potentiodynamic polarization in a 3wt% NaCl solution after preliminary holding of specimens in the medium for 20 minutes for the open circuit potential to stabilize using a potentiostat-galvanostat Elins-P45X. Corrosion characteristics were measured on a three-electrode electrochemical cell with a silver-chloride reference electrode and a platinum counterelectrode. All measurements were performed multiple times (at least three times).

3. Results

The microstructures of Mg-2.2Nd-0.6Zr-0.4Zn (ML10), Mg-6.8RE-2.5Zn-0.6Zr and Mg-8.3RE-2.5Zn-0.6Zr alloys before and after heat treatment are shown in figure 1. In the initial state, all alloys are characterized by a non-equilibrium fine-grained structure. After heat treatment, the alloy structure becomes more homogeneous, and a large number of dispersive inclusions of the second phase (in ML10) or formation of a strengthening LPSO phase occurs. For the ML10 alloy, the dispersive phase was identified as $Mg_{12}Nd$, and for Mg-6.8RE-2.5Zn-0.6Zr and Mg-8.3RE-2.5Zn-0.6Zr alloys, formation of the LPSO phase $Mg_{12}YZn$, eutectic phase $Mg_3Zn_3Y_2$, and insignificant amount of dispersive phase $Mg_{24}Y_5$ was identified. Simultaneously, all alloys are characterized by a slight increase in the average grain size after the heat treatment (figure 1). As a result of heat treatment, a significant improvement in the mechanical properties of the alloys was achieved (table 1), namely, yield strength, σ_T, and tensile strength, σ_B, are increased, while the duration of the aging in the range of 6-24 hours had almost no effect on them.

However, the heat treatment hardening caused by precipitation of dispersive inclusions or LPSO phases resulted, at the same time, in significant decrease of corrosion resistance of the investigated alloys (by 10 times or more). To resolve this issue, these alloys were subjected to MAO, and the effect of such treatment was evaluated. On all alloys, the thickness of oxide layers was about 40 microns, with a relatively small variation for all alloys – the dispersion coefficient is 10-15%. Concurrently, all the obtained oxide layers are characterized by a pronounced two-layer structure: an outer porous layer with separate craters and transverse cracks, and an inner less defective and denser layer directly adjacent to the magnesium substrate (figure 2).

The dominant phase of all oxide layers is magnesium oxide MgO (periclase mineral) produced as a result of direct oxidation of the treated substrate by micro-arc discharges which is similar to [6]. Magnesium oxide content in the layer for all specimens is 98-99 wt%. The magnesium aluminate phase – $MgAl_2O_4$ with the content of 1-2 wt% was also detected. All specimens are characterized by a sufficiently high crystallinity degree – 83-92 vol.% which is in good agreement with [7]. Concurrently, it should be noted that phosphorus and silicon were found in the oxide layers in the amount of 3-5 wt% and 1.8-2.8 wt%, respectively, which are not part of the detected crystalline phases. Apparently, these elements are part of the amorphous phase formed in the oxide layers [6].

The PEO treatment of alloys allowed to significantly increase their corrosion resistance. It was found that after PEO, corrosion current densities decreased by 1-3 orders of magnitude - from 14-19 μA/cm^2 in the initial alloys to 0.016-0.7 μA/cm^2 in the alloys after PEO. The Mg alloy with neodymium experienced the least effect of PEO treatment – a decrease in the corrosion current density by ~ 22 times. For Mg-6.8RE-2.5Zn-0.6Zr and Mg-8.3RE-2.5Zn-0.6Zr (with Gd) alloys, the effect was much more significant - 240-1200 times. Besides, there is no obvious corrosion potential offset as a result of PEO treatment compared to alloys in the initial state.

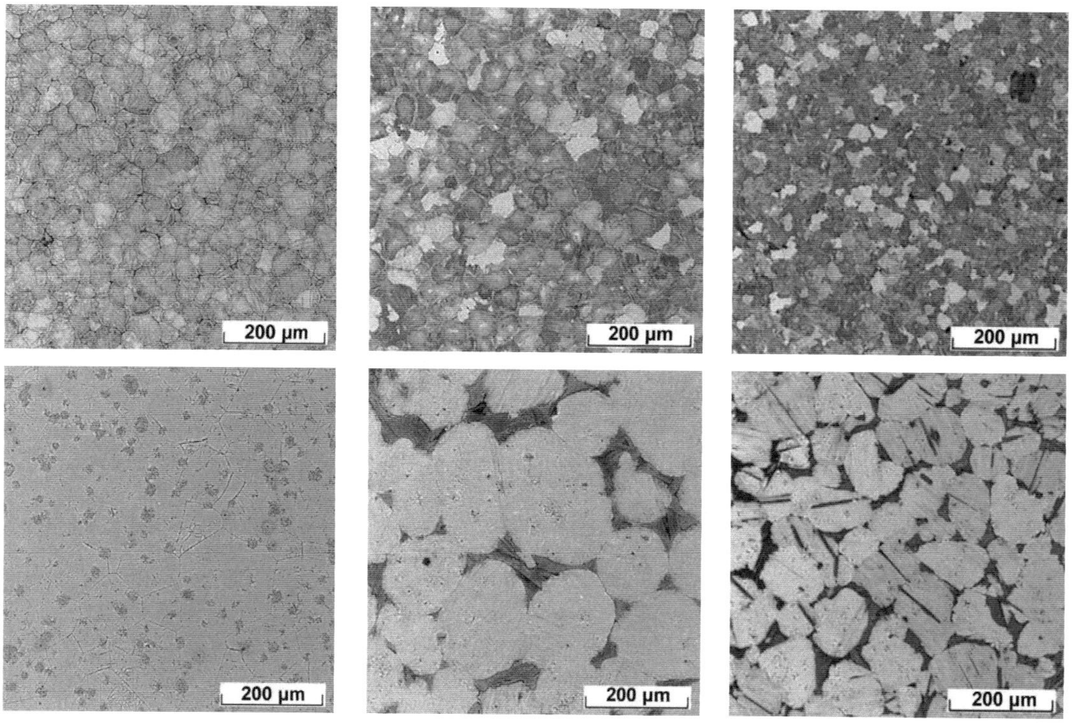

Figure 1. Microstructures of the alloys (from left to right) Mg-2.2Nd-0.6Zr-0.4Zn (ML10), Mg-6.8RE-2.5Zn-0.6Zr and Mg-8.3RE-2.5Zn-0.6Zr before (top row), and after (bottom row) heat treatment in the mode: homogenization 540±10°C, 12 h + air quenching + aging 200±5°C, 12 h.

Table 1. Mechanical properties of alloys Mg-2.2Nd-0.6Zr-0.4Zn (ML10), Mg-6.8RE-2.5Zn-0.6Zr and Mg-8.3RE-2.5Zn-0.6Zr before and after heat treatment in the T6 mode.

Alloy, state	σ_T, MPa	σ_B, MPa	ε, %	HB
ML10, cast	108	174	6.9	55
ML10(T6) (GOST 2856-79)	140	238	3	70.5
Mg-6.8RE-2.5Zn-0.6Zr, cast	126.7	197.3	6.2	63.5
Mg-6.8RE-2.5Zn-0.6Zr (T6)	132.6	241	7.9	69.5
Mg-8.3RE-2.5Zn-0.6Zr, cast	133	201	4.0	66.1
Mg-8.3RE-2.5Zn-0.6Zr (T6)	138	259	13.6	74.7

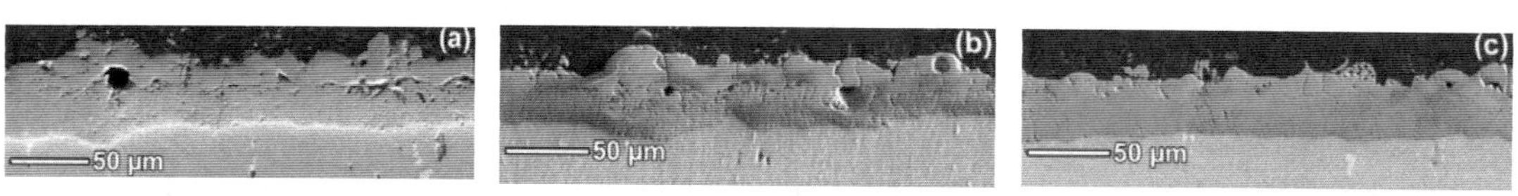

Figure 2. The structure of oxide layers (cross-sections, SEM) on the alloys Mg-2.2Nd-0.6Zr-0.4Zn (a), Mg-6.8RE-2.5Zn-0.6Zr (b) and Mg-8.3RE-2.5Zn-0.6Zr (c).

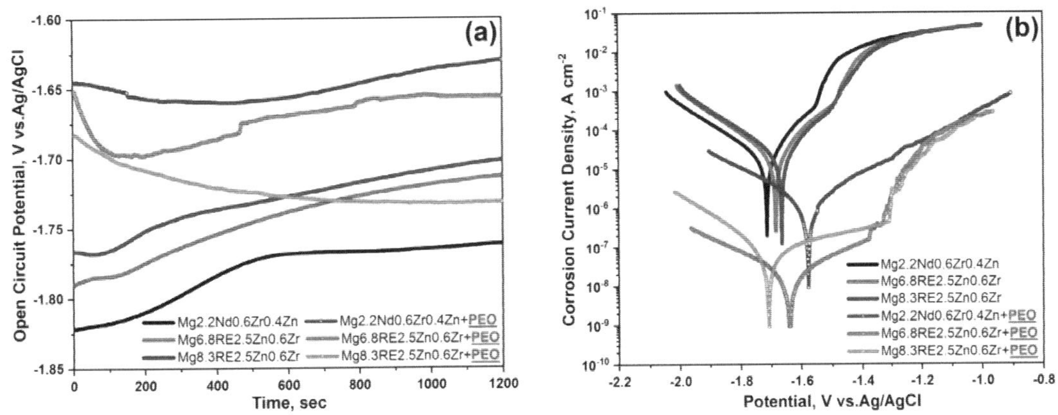

Figure 3. OCP change (a) and Tafel (b) curves for alloy specimens in the initial (bare) state and after surface modification by PEO method in 3.5% NaCl solution.

Table 2. Basic electrochemical corrosion indicators (ASTM G102-89) of the investigated magnesium alloys[a].

Specimen	Ecorr, mV	icorr, μA/cm2	Rp, Ω·cm2	CR, mm/yr	MR, g/(m2·d)
ML10(T6)	-1692±17	16.68±0.34	(1.78±0.04)·103	0.348	1.825
ML10(T6)+PEO	-1575±3	0.732±0.138	(5.60±1.27)·104	1.4·10-2	7.1·10-2
Mg-6.8RE-2.5Zn-0.6Zr(T6)	-1663±21	18.96±2.10	(1.72±0.16)·103	0.354	2.026
Mg-6.8RE-2.5Zn-0.6Zr(T6)+PEO	-1634±5	0.016±0.002	(3.07±0.01)·106	0.3·10-3	1.6·10-3
Mg-8.3RE-2.5Zn-0.6Zr(T6)	-1677±31	14.01±4.15	(1.69±0.47)·103	0.349	2.081
Mg-8.3RE-2.5Zn-0.6Zr(T6)+PEO	-1715±18	0.058±0.021	(1.94±0.82)·106	0.9·10-3	5.2·10-3

[a] ±σ is given as an error

It should be noted that localized pitting corrosion is typical for all the investigated alloys, therefore, calculation of the corrosion rate and mass loss may yield some error and, most likely, the CR and MR values in table 2 are somewhat underestimated [8]. However, comparative tests of alloys can be carried out using the Tafel curves, both between the alloys, and to evaluate the efficiency of treatment of each alloy by PEO method.

4. Conclusions

It can be concluded that the heat treatment of cast magnesium alloys with RE additions results in an increase in yield and tensile strength by ~ 5-20%, and in plasticity by almost 2 times, the effect mainly provided by the presence of the LPSO phase. The Mg-8.3RE-Y2.5Zn0.6Zr alloy after the applied heat treatment has the best mechanical properties among the investigated alloys (elongation ~12% and hardness ~75 HB). In addition, this alloy has additional potential for increasing strength by optimizing heat treatment modes. At the same time, the alloy chemical composition has almost no effect on the oxidability and phase composition of the oxide layers produced by MAO.

Treatment of alloys by MAO method improves the corrosion resistance of the investigated alloys due to formation of the barrier oxide layer. The oxide layer prevents the access of corrosive medium to the substrate metal and by this way prevents the dissolution of metal during prolonged holding in a corrosive medium. This also reduces the corrosion currents (when specimens are polarized) by about 3 orders of magnitude.

Acknowledgements

Research of mechanical properties of experimental Mg alloys was conducted by the courtesy of the Ministry of Science and Higher Education of the Russian Federation as part of the state assignment (Project FEMR-2020-0003). Plasma electrolytic oxidation and electrochemical investigations of specimens were financed by a grant from the Russian Science Foundation (Project No. 21-19-00656). Finally, we would like to express our gratitude to Mr. Anton Polunin and Dr. Alexander Katsman (Togliatti State University) for their help in treating PEO and measuring XRD.

References

[1] Czerwinski F 2012 *Magnesium Alloys - Corrosion and Surface Treatments* (InTech)
[2] Li C Q, Xu D K, Zeng Z R, Wang B J, Sheng L Y, Chen X B and Han E H 2017 *Mater. Des.* **121** 430–41
[3] Sikdar S, Menezes P V, Maccione R, Jacob T and Menezes P L 2021 *Nanomaterials* **11** 1375
[4] Lu X, Mohedano M, Blawert C, Matykina E, Arrabal R, Kainer K U and Zheludkevich M L 2016 *Surf. Coatings Technol.* **307** 1165–82
[5] Polunin A V, Cheretaeva A O, Borgardt E D, Shafeev M R, Katsman A V and Krishtal M M 2020 *J. Phys. Conf. Ser.* **1713** 12036
[6] Lu X, Blawert C, Huang Y, Ovri H, Zheludkevich M L and Kainer K U 2016 *Electrochim. Acta* **187** 20–33
[7] Curran J A and Clyne T W 2005 *Surf. Coatings Technol.* **199** 177–83
[8] Shi Z, Cao F, Song G L and Atrens A 2014 *Corros. Sci.* **88** 434–43

Residual stresses in surface layers and coatings

A A Ashmarin[1], A A Lozovan[2], S Ya Betsofen[2], V S Moiseyev[2], E P Kubatina[2] and M A Lebedev[2]

[1] IMET RAN, Moscow, 117334, Russia
[2] Moscow Aviation Institute (NRU), 125993, Moscow, Russia

E-mail: s.betsofen@gmail.com

Abstract. Methods of measurement and mechanisms of formation of residual stresses in the surface layers of metal alloys after ion nitriding, as well as in heat-shielding and refractory coatings are considered. Examples of the application of a technique based on the peculiarities of elastic anisotropy for measuring residual stresses in heterogeneous surface layers, for which the standard technique of diffraction tensometry "$\sin^2\Psi$" is not applicable, are given.

1. Introduction

Residual stresses play an important role in the formation of modified surface layers. However, the experimental determination of these stresses encounters fundamental difficulties. This is due to the fact that the deformation of the lattice during the formation of an interstitial and substitutional solid solution and from elastic stresses gives a similar diffraction effect. Therefore, it is necessary to develop methods for separating the contribution of residual stresses and heterogeneity of the chemical composition in surface layers and coatings.

2. Methods for measuring residual stresses

The $\sin^2\Psi$ method is widely used in diffraction tensometry, using both X-ray [1, 2] and neutron radiation [3, 4]. When determining the residual stresses by the "$\sin^2\Psi$" method, X-ray is used, including with oblique incidence of the beam, which leads to errors in the presence of a gradient of the lattice parameters along the depth, since the thickness of the scattering layer is less than in symmetric X-ray. In [5, 6], a method was developed for separating the effects of stresses and the gradient of lattice parameters, based on the features of elastic anisotropy of crystals. It is shown that to estimate the residual stresses, one can analyze the lattice parameters of grains in which different reflection planes are parallel to the sample surface, while for any pair of reflections ($h_1k_1l_1$) and ($h_2k_2l_2$), having different values of elastic constants ($K_{hkl}=(\nu/E)_{hkl}$) the value of residual stresses is determined from the ratio:

$$\sigma_{ocm.} = \frac{a_{h_1k_1l_1} - a_{h_2k_2l_2}}{2(a_{h_2k_2l_2}K_{h_1k_1l_1} - a_{h_1k_1l_1}K_{h_2k_2l_2})} \quad (1)$$

Figure 1 shows the lattice parameters calculated from (1) for various reflections for α- and γ-Fe for $\sigma = \pm$ 500 MPa and a_o = 2.90 Å and 3.60 Å, respectively, for α- and γ-Fe.

Content from this work may be used under the terms of the Creative Commons Attribution 3.0 licence. Any further distribution of this work must maintain attribution to the author(s) and the title of the work, journal citation and DOI.
Published under licence by IOP Publishing Ltd

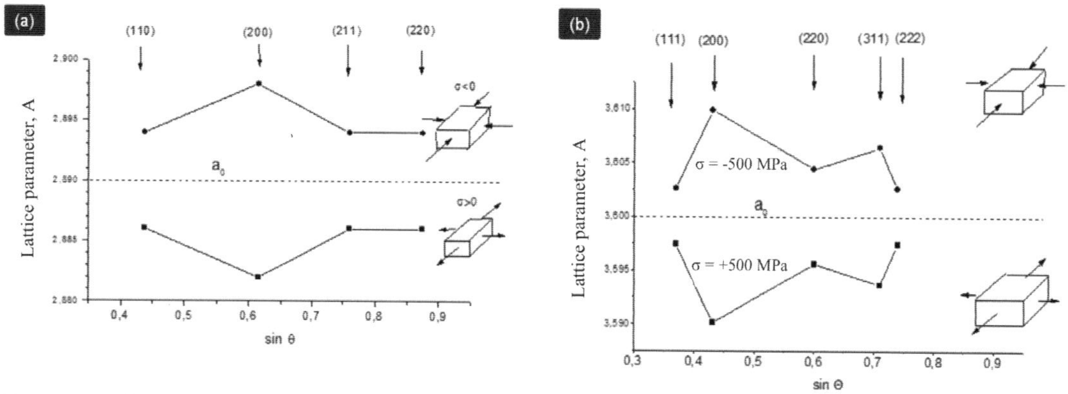

Figure 1. The lattice parameters (hkl) of reflections calculated from Eq. (1) for residual compressive and tensile stresses in the α-phase (a) and γ-phase (b) iron [8, 9].

3. Residual stresses in surface layers and coatings

3.1. Residual stresses in thermal barrier coatings (TBC)

The service life of a zirconium oxide TBC, partially stabilized with 8 wt% yttrium oxide, is largely determined by the level of residual stresses, which reach especially high values in the region of the ceramic/metal sublayer interface, where TGO is formed (thermally growing oxide - the oxidation product of the metal sublayer, consisting mainly of aluminum oxide) [7, 8]. In [9], the distribution of residual stresses in the TBC was investigated using an original X-ray technique based on the anisotropy of the elastic properties of the tetragonal modification

In figure 2 shows the distribution of residual stresses over the section of the TBC.

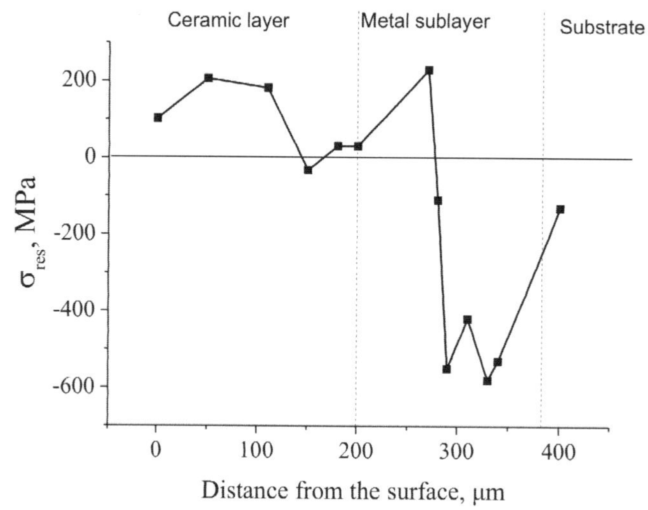

Figure 2. Distribution of residual stresses over the thickness of the TBC [11].

An increase in the specific volume of the metal sublayer as a result of the formation of TGO leads to the formation of high compressive stresses in this sublayer, reaching 600 MPa. At the same time, tensile stresses up to 200 MPa are present in the ceramic layer, which are probably of quenching origin. Despite the fact that the magnitude of the tensile stresses is not large, it is essential that they are localized in the outer ceramic layer and therefore pose a hazard.

3.2. Residual stresses in mono- and multilayer coatings

The deposition of multilayer coatings is one of the promising technologies for surface modification [10–12] due to their advantages over monolayer coatings such as improved adhesion of the coating to the substrate; expanding the possibility of optimizing a set of properties by applying intermediate layers with different mechanical characteristics, the efficiency of interfaces for stress relaxation and retardation of crack propagation.

In [13–16], the formation of texture and residual stresses in monolayer Ta, W, Mo, Nb magnetron coatings was investigated depending on the bias voltage on a copper substrate from 0 to -300 V. All coatings are characterized by high compressive stresses reaching 2000 MPa (figure 3), due to a significant difference in the LTEC values of refractory metals and Cu, while the values of residual stresses increase with increasing bias voltage on the substrate.

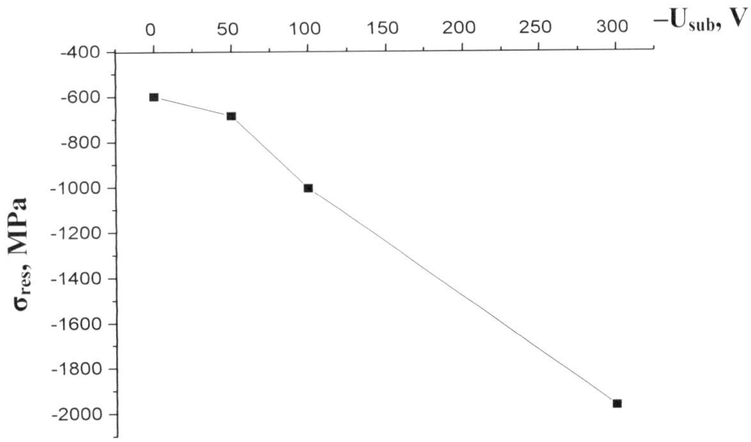

Figure 3. Dependence of the magnitude of residual stresses in W magnetron coatings on the voltage on the Cu substrate [14].

For a multilayer (n = 120) Mo-Nb coating [16] ~ 800 μm thick on a Ni-Cr substrate (figure 4a), relatively low residual compressive stresses were found, amounting to -576 MPa in the subsurface Mo layer and -213 MPa in the outer Nb layer. layer, which is much lower than in monolayer coatings of these metals with a thickness of ~ 10 μm.

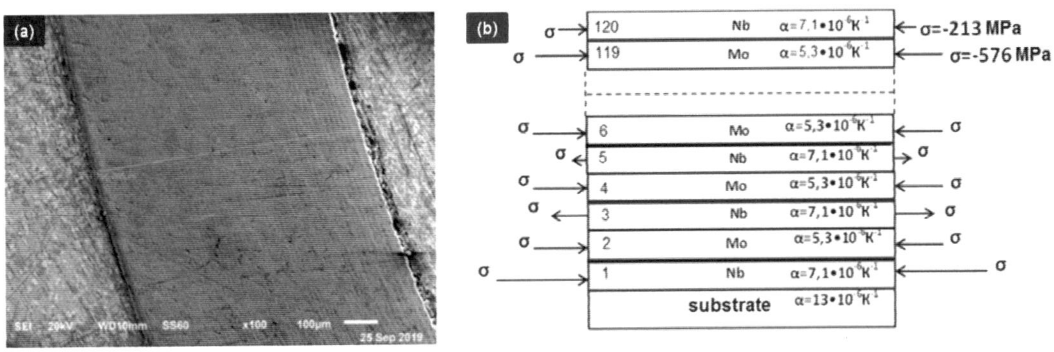

Figure 4. Nb-Mo cross section (a) and (b) stress state formation diagram in a multilayer magnetron coating [16].

To explain the low residual stresses in multilayer Nb-Mo coatings, a mechanism is proposed, which is demonstrated by the diagram in figure 4b. The main reason for low stresses is that the alternation of layers of refractory metals differing in the LTEC values (α_{Mo} = 5.3 10^{-6} and α_{Nb} = 7.1 $10^{-6}K^{-1}$) leads to mutual compensation of the thermal component of stresses. The first Nb layer on a chromium-nickel substrate, characterized by a significantly higher LTEC (α_{Ni-Cr} = 13 $10^{-6}K^{-1}$), experiences compressive stresses (figure 4b). The next Mo layer with an LTEC value lower than Nb also experiences compressive stresses, i.e. the total level of compressive stresses increases, but the next Nb layer with a large LTEC value in comparison with Mo already experiences tensile stresses, which compensates for the component of compressive stresses in the previous layers (figure 4b).

4. Conclusion

The efficiency of the application of the original X-ray technique for determining residual macrostresses by the X-ray method, which can be effectively used for coatings and surface layers with a gradient structure, is shown. The technique is based on symmetric X-ray photography of reflections from atomic planes with the maximum difference in elastic characteristics and calculation of stresses from the difference in elastic deformation of these planes.

References

[1] Jing Zheng 2018 *Procedia Engineering* **213** 36–47

[2] Lin Chen, Gaoyang Mi, Xiong Zhang and Chunming Wang 2019 *Materials and Design* **168** 107653

[3] Jeong-ung Park, Gyubaek An and Wanchuck Woo 2018 *International Journal of Naval Architecture and Ocean Engineering* **10** 129–40

[4] Javadi Y, Smith M C, Abburi K et al. 2017 *International Journal of Pressure Vessels and Piping* **154** 41–57

[5] Betsofen S Ya, Petrov L M, Ilyin A A, Bannykh I O and Lutsenko A N 2004 *Surface. X-ray, synchrotron and neutron research* **1** 39–45

[6] Jeong-ung Park, Gyubaek An and Wanchuck Woo 2018 *International Journal of Naval Architecture and Ocean Engineering* **10** 129–40

[7] Kartal M E, Kang Y-H, Korsunsky A M, Cocks A C F and Bouchard J P 2016 *International Journal of Solids and Structures* **80** 420–9

[8] Martinson P, Daneshpour S, Koçak M, Riekehr S and Staron P 2009 *Mater Design* **30(9)** 3351–9

[9] Betsofen S Ya, Lozovan A A, Lenkovets A S, Labutin A A and Grushin I A 2021 *Metals* **4** 87–98

[10] Ahrens M, VaЯen R and Stoëver D 2002 *Surface and Coatings Technology* **161** 26–35

[11] Evans H E 2010 *Shreir's Corrosion, Volume 1: Basic Concepts, High Temperature Corrosion* 691–724

[12] Thobor A, Rousselot C, Clement C, Takadoum J, Martin N, Sanjines R and Levy F 2000 *Surface and Coatings Technology* **124** 210–21

[13] Andersen K N, Bienk E J, Schweitz K O, Reitz H and Chevallier J *Surface Coating and Technology* **123** 219–26

[14] Li T S, Li H and Pan F 2001 *Surface Coating and Technology* **137** 225–9

Research of the influence of bias voltage on the structure and residual stress in Ta/W coatings applied on a copper substrate by inverted magnetron

A A Lozovan, A S Lenkovets, N A Ivanov and T S Sukhova

Moscow Aviation Institute (NRU), 125993, Moscow, Russia

E-mail: loz-plasma@yandex.ru

Abstract. This paper presents the results of studying the effect of the bias voltage on a copper tube substrate on the texture and residual stresses in 4-layer Ta/W/Ta/W coatings deposited with an inverted magnetron. It is shown that, in contrast to monolayer coatings of Ta and W, in which residual stresses exceeding 2 GPa are formed on the substrate at high voltages on the substrate, stress relaxations occur in the 4-layer coating in alternating layers differing in LTEC values and in the outer W-layer of stress practically absent.

Thin films and coatings made of tungsten are of considerable interest for use in various fields: microelectronics [1] (including spintronics [2, 3]), thermo photovoltaic converters in power engineering, high-temperature nanophotonics [4] or others. One of the possible areas of application W coatings are thermal barrier coatings for parts of future thermonuclear reactors, such as ITER [5], which will be exposed to extreme thermal loads and ion bombardment, as well as for other thermally loaded products, such as combustion chambers of rocket engines. The key problems for such coatings are thermo mechanical stability with respect to delamination, oxide formation and diffusion at high operating temperatures, and the achievement of long-term high-temperature grain stability [4].

For deposition of W, pulsed laser deposition, atomic layer deposition, and sputtering by planar DC magnetrons are commonly used. In [5, 6], we showed that for the creation of thin-walled small-sized axis symmetric shell structures from layered composites, for example, pipe products with different surface profiles, a system of sequentially located inverted magnetrons and one ordinary cylindrical magnetron used for cleaning substrates is very effective. The system allows the formation of layered composite shells by spraying various layers onto a mandrel (for example, copper), which is subsequently etched away.

It is known that the deposition of W-coatings on Cu substrates, which have a significantly higher thermal coefficient of linear expansion (TCLE), leads to huge residual stresses, reaching -5 GPa. The magnitude of these stresses, in addition to the difference in the LTEC and Young's modulus, depend on the temperature of the coating process, which is determined by the energy parameters of the deposition process, primarily the voltage on the substrate. In our works [7, 8], we investigated the effect of voltage on the substrate on the formation of texture and residual stresses in monolayer W and Ta coatings deposited on a Cu substrate by an inverted magnetron. In our works [7, 8], we studied the effect of voltage on the substrate on the formation of texture and residual stresses in monolayer W and Ta coatings deposited on a Cu substrate by an inverted magnetron. At high voltages from -100 to -300 V, high compressive residual stresses exceeding 2 GPa were found in the coatings, which led to

Content from this work may be used under the terms of the Creative Commons Attribution 3.0 licence. Any further distribution of this work must maintain attribution to the author(s) and the title of the work, journal citation and DOI.
Published under licence by IOP Publishing Ltd

disruption of adhesion and destruction of coatings. In [9], when studying the formation of residual stresses in monolayer coatings of Mo and Nb, high residual compressive stresses were also found. However, in a multilayer (n = 120) Mo / Nb coating 800 μm thick, the stresses in the outer layers did not exceed 600 MPa. In this regard, in this work, we investigated the effect of the voltage on the substrate on the residual stresses, as well as on the texture and crystal lattice distortions of coatings of 4-layer Ta/W/Ta W coatings deposited using an inverted magnetron.

Spraying was carried out on a specialized installation MRM-1, presented in [6]. Argon with a purity of at least 99.9% was used as a working gas. The cathode material is W purity ≈ 99.9%, the inner diameter and length of the cathode were 37 mm and 24 mm. A copper tube M-1 with a diameter of 10 mm and a length of 20 mm was used as a substrate. Before sputtering, the tube was polished and degreased. Then the substrate was installed on the rod of vertical movement of the samples in the chamber and the chamber of the setup was evacuated to a residual pressure of 10^{-3} Pa. Before spraying, for 30 minutes, cleaning with a glow discharge plasma was carried out at an argon pressure of 5 Pa and a voltage on the substrate of 1100 V. Then, tungsten was deposited at various bias voltages on the substrate according to the modes shown in table 1. The cathode axis and periodically completely left the cathode region, alternately completely leaving its ends. The total spraying time for all samples is 8 hours. Each layer was applied for 2 hours. alternation of layers: Ta-W-Ta-W.

Table 1. Spray mode No.1 (bias voltage 0 V).

Layer	U_{m3}, V	I_{m3}, A	P_{Ar}, Pa	T, °C
Ta	280-285	1	0,2	420
W	290-305	1	0,2	430
Ta	275-285	1	0,2	415
W	290-305	1	0,2	430

Table 2. Spraying mode No.2-4 (bias voltage -50, -100, -150, -200V).

Layer	U_{m3}, V	I_{m3}, A	-Us, V	Is, A	P_{Ar}, Pa	T, °C
Ta	270-280	1	50 (100, 150,200)	0,14-0,05	0,2	430
W	290-300	1	50 (100, 150,200)	0,14-0,05	0,2	440
Ta	270-285	1	50 (100, 150,200)	0,14-0,05	0,2	430
W	290-305	1	50 (100, 150,200)	0,14-0,05	0,2	440

Inverse pole figures (IPFs) were obtained by taking X-ray diffraction patterns in the angular range 2θ = 30–140°. The pole density of 6 independent reflections hkl on the stereographic triangle: 001, 011, 013, 111, 112, 123, (figure 1a) was determined from the ratio:

$$P_{hkl} = \frac{n\,(I_{hkl}^{\text{тек}} / I_{hkl}^{\text{эт}})_i}{\sum_{i=1}^{n} (I_{hkl}^{\text{тек}} / I_{hkl}^{\text{эт}})_i} \tag{1}$$

where, $I_{hkl}^{\text{тек}}$, $I_{hkl}^{\text{эт}}$ are the integral intensities of the hkl reflections for the textured and textureless (reference) sample, respectively; n is the number of independent hkl reflections (n = 6).

Figure 1 shows the IPF for W coatings deposited on a copper substrate at bias voltages on the substrate from 0 to -200 V. It can be seen that with an increase in the substrate voltage, changes in the coating texture occur, which are of a non-monotonic nature.

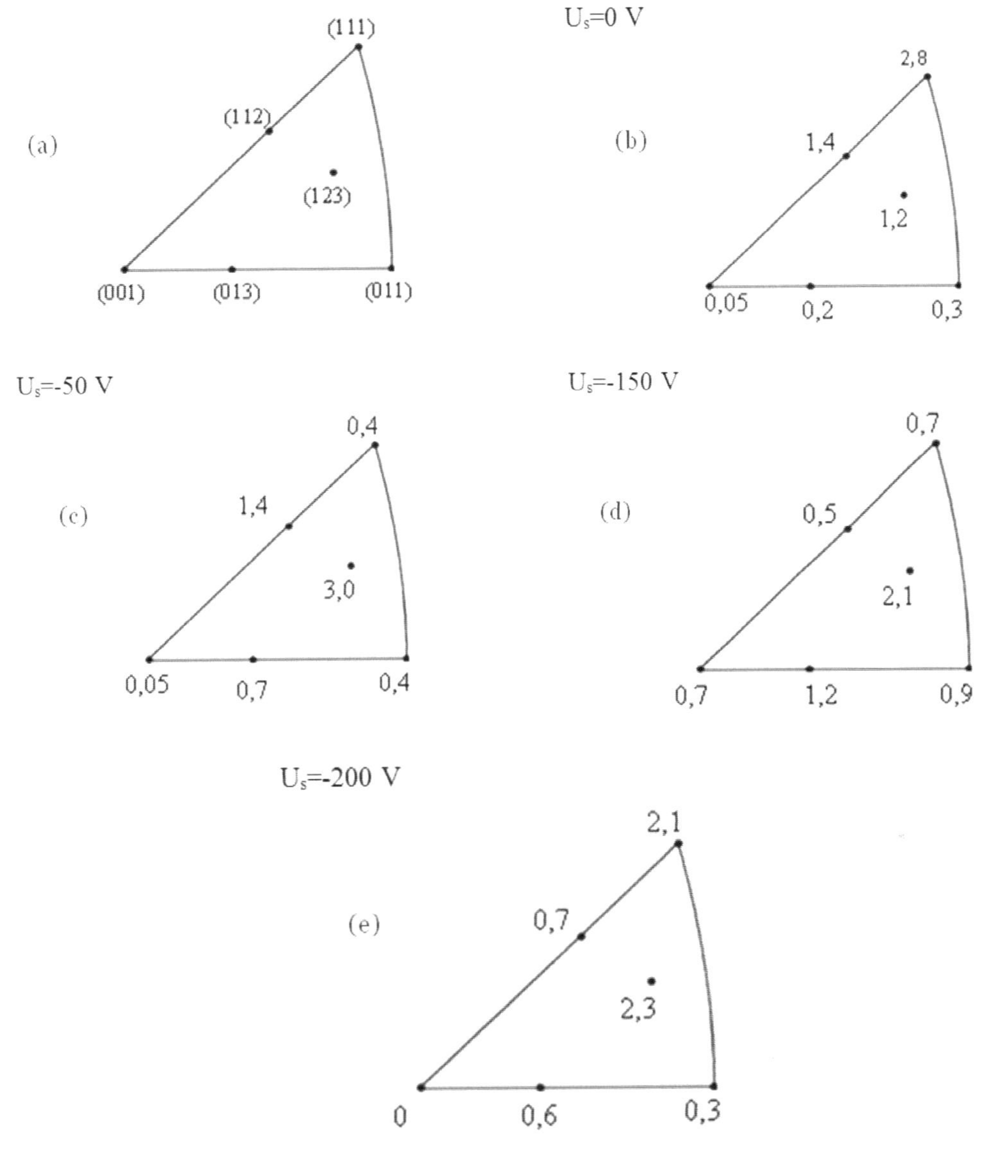

Figure 1. IPF for the outer W layer of 4-layers coatings deposited at different voltage values on the substrate: (a) a standard stereographic triangle for a cubic lattice; (b) Us = 0 V; (c) Us = -50 V; (d) Us= -150 V; (e) Us = -200 V.

In figure 2, these results are summarized in the form of the dependence of the pole density (111) on the voltage on the substrate. Figure 3 shows the voltage dependences of the half-widths of the diffraction lines (222) and (321). It can be seen that with an increase in the voltage on the substrate, the width of the diffraction lines increases, which indicates an increase in the distortion of the crystal lattice and a refinement of the subgrain structure. Figure 4a shows the values of the lattice parameters of the coatings calculated from different interplanar distances (hkl). The presented results indicate the absence of noticeable residual stresses in the coatings, which distinguishes them from monolayer coatings, in which stresses exceeding 2 GPa were found.

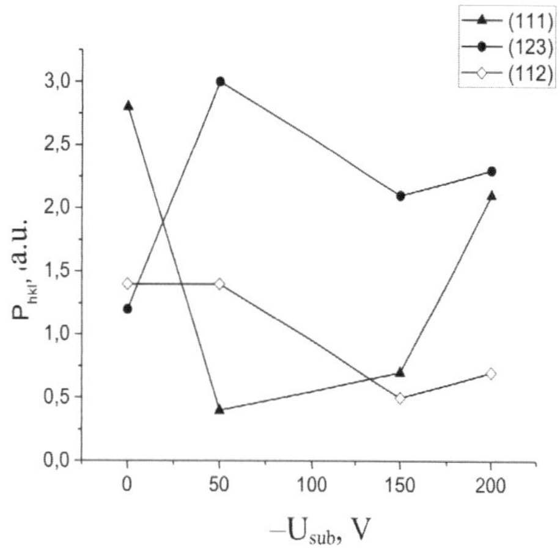

Figure 2. Pole density for orientation (111), (123), and (112) versus substrate voltage.

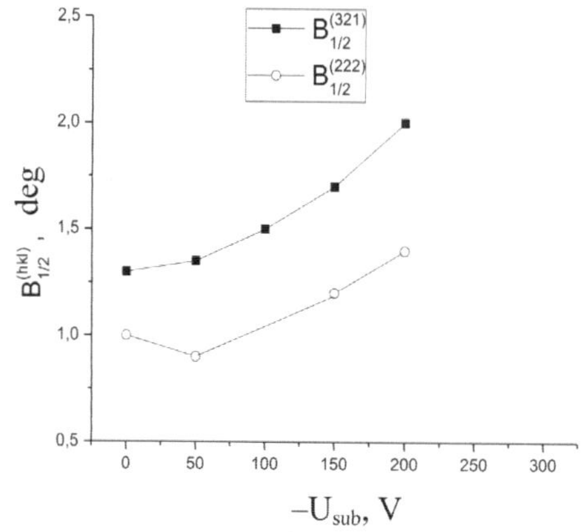

Figure 3. Dependence of the half-widths of the diffraction lines (321) and (222) on the voltage on the substrate.

It is likely that stress relaxation occurred in the 4-layer coating due to mutual compensation of thermal stresses in multilayer coatings, when the compressive stresses arising when a Ta layer is deposited on a W sublayer with a lower LTEC are balanced by tensile stresses when the next W layer is deposited on a Ta sublayer with higher LTEC, figure 4b.

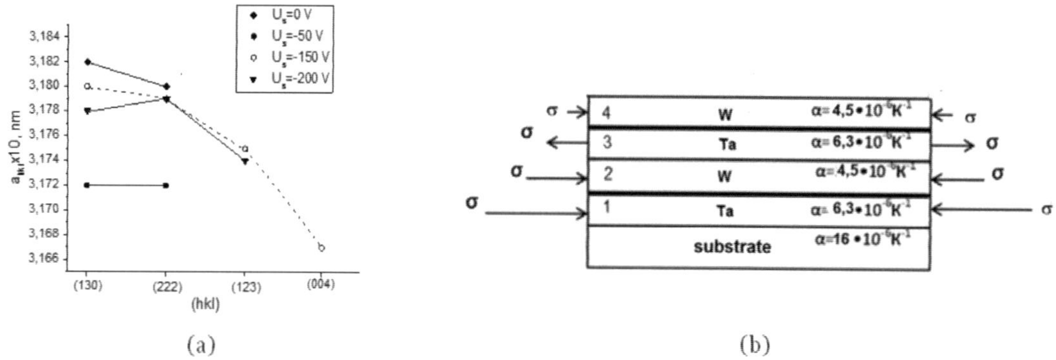

Figure 4. Coating lattice parameters calculated for different interplanar distances (hkl) (a) and scheme of formation of residual stresses in a 4-layer Ta/W/Ta/W coating.

A similar situation was observed in a 120-layer Nb/Mo coating with a thickness of 800 μm [9], in which the compressive stresses were <600 MPa, while in monolayer 10 μm coatings made of the same metals the values of the compressive stresses on the copper substrate exceeded 2000 MPa. The increased lattice periods (figure 4a) of the coatings with respect to pure W (a = 0.317 nm) are possibly due to the dissolution of Ta atoms with a higher period in the W lattice (a = 0.33 nm). The lattice period of 0.318 nm corresponds to the dissolution of 5-7 at.% Ta.

1. With an increase in the voltage on the substrate, the width of the diffraction lines increases, which indicates an increase in the distortion of the crystal lattice and the refinement of the subgrain structure.

2. With an increase in the voltage on the substrate, changes in the texture of the coatings occur, which are of a non-monotonic character.

3. No noticeable residual stresses were found in 4-layer Ta/W/Ta/W coatings, which distinguishes them from monolayer coatings and is due to mutual compensation of thermal stresses in successive Ta and W layers with different LTEC values (6.3 and 4.5 10^{-6} deg^{-1} respectively).

Acknowledgments
The authors are grateful to S. Ya.Betsofen for a discussion of the results of X-ray studies and valuable comments.

References

[1] Choi D B, Wang S, Chung et al. 2011 *J. Vac. Sci. Technol.* **A 29** 051512
[2] Pai C F, Liu L, Li Y, Tseng H W, Ralph D C and Buhrman R A 2012 *Appl. Phys. Lett.* **101** 122404
[3] Vüllers F T N and Spolenak R 2015 *Thin Solid Films* **577** 26
[4] Stelmakh V, Rinnerbauer V, Joannopoulos J D et al. 2013 *J. Vac. Sci. Technol.* **A31** 061505
[5] Kapp M, Martinschitz K J, Keckes J et al. 2008 *BHM Berg- und üttenmännische Monatshefte* **153** 273–5
[6] Todoran A, Mantel M, Bes A et al. 2014 *Plasma Sources Sci. Technol.* **23** 065039–7
[7] Lozovan A A, Lenkovets A S, Ivanov N A, Alexandrova S S and Kubatina E P 2018 *Mater. XV-th International scientific-technical conference* (October 16-17, 2018 MAI: Space-2000 pp 152–6
[8] Betsofen S Ya, Lozovan A A, Lenkovets A S, Labutin A A and Grushin I A 2021 Metals **4** 87–98

Optical fluxes propagation in the planar transport layer of multilayer coatings

E V Egorov[1,2,3] and V K Egorov[1]

[1] Institute of Microelectronics Technology RAS, 142432, Chernogolovka, Russia
[2] Institute of Radio Engineering and Electronics RAS, 141190, Fryazino, Russia
[3] Financial University under the government of the Russian Federation, Moscow, Russia

E-mail: egorov@iptm.ru

Abstract. The article is concerned with peculiarities study of the quasimonochromatic optical fluxes propagation through thin planar transparent layer of multilayer coating. There is shown that these fluxes can be transported by the layer in process of its multiple consequtive total internal reflection or by the waveguide-resonance propagation manner depending on correlation between the layer width and the radiation coherence length half of transported fluxes. Efficiency comparison of these radiation transportation mechanisms showed that the waveguide-resonance propagation approach is more adequate for results description of the optical waveguides functioning. It allowed to conclude that optical waveguides (fibers) function in frame of the waveguide-resonance paradigm and the waveguide-resonance mechanism is responsible for the light fluxes transportation on great distances.

1. Introduction

One of the most urgent tasks connected with study of planar thin-film coatings properties is the phenomena investigation of the different radiation fluxes interaction with the coatings ingredients. The most attention of these investigations is oriented on the interactions analysis, which take place in conditions when coatings in gradient thicknesses is commensurate with sizes of fluxes radiation parameters. There are the radiation wavelength (λ_0) the radiation monochromatization degree ($\Delta\lambda$) the radiation coherence length ($L=\lambda_0^2/\Delta\lambda$) and radiation standing wave interference field parameters appeared in the interaction result of incident and reflected fluxes (sizes of fields and standing wave period). Similar investigations of X-ray quasimonochromatic fluxes interaction with thin film structures allowed to fix the specific, early unknown mechanism of radiation fluxes propagation – its waveguide-resonance streaming and conditions, which are necessary for the mechanism realization [1]. Moreover, on basis of this mechanism, it was possible to build the planar X-ray waveguide-resonator (PXWR) – device, which can form X-ray beams with nanosize cross-section and enhanced radiation density [2]. Waveguide-resonance devices were built by use the phenomenon of quasimonochromatic X-ray fluxes total external reflection on the material interface [3]. Analogical devices in very limited quantities were constructed in frame the Bragg reflection application [4]. Its got name the Bragg-Laue waveguide-resonance cells (BLWRC). In principle, it is possible to built similar devices by use the mirror light reflection on metallic coatings [5], and the total internal reflection phenomenon (TIR) featured for the interface between optically transparent mediums with

Content from this work may be used under the terms of the Creative Commons Attribution 3.0 licence. Any further distribution of this work must maintain attribution to the author(s) and the title of the work, journal citation and DOI.
Published under licence by IOP Publishing Ltd

different refraction index [6,7]. At the same time, it is essential that the TIR phenomenon take place when the initial light flux propagates in the medium with more high refraction index. This phenomenon can be used for the optical radiation fluxes transportation by multilayer coating with optical transparent layers, presented on figure 1. Similar structure has name of the symmetrical optical waveguide [8–10].

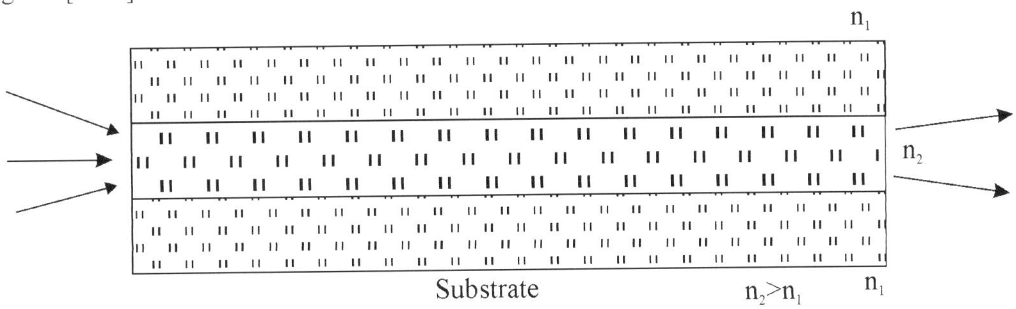

Figure 1. Sample of the double layer coating on transparent substrate for the planar symmetrical optical waveguide formation built on base of the total internal reflection phenomenon.

2.Phenomenon of light total internal reflection

Flux of the optical radiation is characterized by the average wavelength (λ_0), the magnitude of its monochromatism degree ($\Delta\lambda$) and the value of radiation coherence length ($L=\lambda_0^2/\Delta\lambda$), which is reflected the light photon longitudinal size [11]. If the flux propagates in the transparent medium featured by n_2 refraction index and incidences on the planar interface with the transparent medium featured by n_1 refraction index in conditions of $n_2>n_1$ we can observe the flux total internal reflection phenomenon [7]. Owing to the coherence of incident and reflected fluxes the interaction arised. In the result, in stationary conditions the optical standing wave interference field area appears [12]. Scheme of it arising is presented on figure 2.

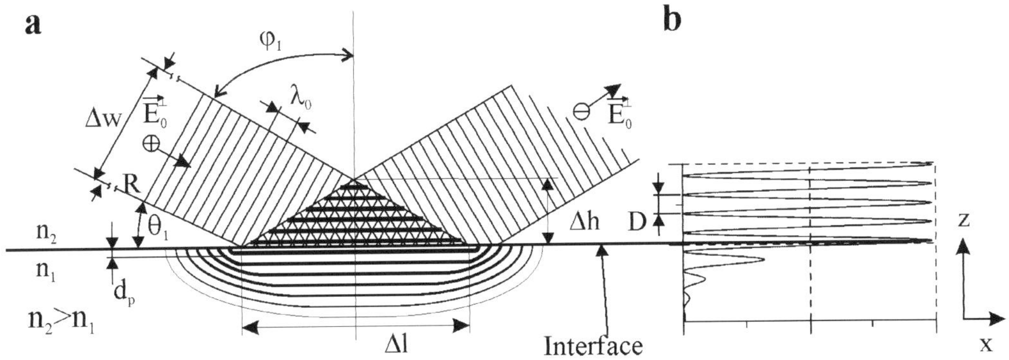

Figure 2. Scheme of the radiation standing wave interference field area appearing at total internal reflection of the quasimonochromatic optical radiation flux on planar interface between transparent mediums with refraction indexes n_1 and n_2 ($n_2>n_1$). Scheme does not take account the Goos-Hanchen shift. d_p – is the interference field penetration depth, D – is the standing wave period, φ_1 is the radiation flux incident angle exceeding the critical total reflection one, Δw – is the incident flux width.

It is a need to notices that the size of interference field area in the medium with n_2 refraction index will have a limited value whereas the interference area in the medium with smaller refraction index will be unrestricted. However, the interference field intensity in it will be decreased exponentially as distance increasing from the interface. The optical standing wave period (D_i) in different mediums will be defined by the expression:

$$D_i = \frac{\lambda_0}{2n_i \cos\varphi}; (i = 1, 2) \quad (1)$$

where φ_2 is the flux incident angle. The interference field propagation in the medium featured by n_1 refraction index is usually characterized by the penetration depth parameter (d_p), which indicates the material surface layer thickness corresponding to the interference field intensity decreasing on "e" times. In conditions of the light flux total internal reflection the parameter is defined by the expression [13].

$$d_p = \frac{\lambda_0}{2\pi\sqrt{\sin^2\varphi_1 - \left(n_1/n_2\right)^2}}; \left(\varphi_1 > \arcsin\frac{n_1}{n_2}\right) \quad (2)$$

It is very important to evaluate the interference field area size in the medium with more higher refraction index n_2. The size limitation of this area is connected with the interference interaction peculiarity. The interference is possible when the phase shift between coherence optical fluxes will be smaller as 180° degrees [12]. At the first glance seems that the area size is defined by cross-section of the incidence optical flux (Δw). But from the physical points of view, it size is restricted by the radiation monochromatization degree or the phase criterion [12]:

$$\Delta\varphi \leq 2\pi \lambda_0^2/2\Delta\lambda \quad (3)$$

The criterion can be transformed into the linear form:

$$\Delta l = \frac{\lambda_0}{2\pi}\Delta\varphi = \lambda_0^2/2\Delta\lambda = L/2 \quad (4)$$

where L is the radiation coherence length. By these means, peculiarities of the interference field appearing define its longitudinal magnitude. The evaluation of interference field transverse size (Δh) demands the specific consideration. On base of the formal approach in frame of the scheme presented on figure 2 one can expected that the transverse parameter size will be correlated with the dimension range:

$$0 \leq \Delta h \leq \frac{L\sin\varphi_{crit}}{4} \quad (5)$$

where φ_{crit} is the TIR critical angle. However, this approach ignores the fundamental principle of the interference phenomenon – the conception of radiation fluxes spatial coherence [14]. The conception foresees the interference possibilities of photon on oneself and the different photons at conditions when its were imitated from the same place of the radiation source [15]. By the conception use, in the wake of the fundamental work results [14] one can expected that transverse and longitudinal magnitudes of the local interference field area formed in the medium with larger refraction index in conditions of the quasimonochromatic optical flux total internal reflection will be equivalented and will be defined by the radiation coherence length half (L/2). Analogical conclusion was produced at consideration of the total external reflection phenomenon of quasimonochromatic X-ray radiation fluxes on the material interface [2].

3. Propagation peculiarities of optical fluxes in the planar transparent layer

The TIR phenomenon is the fundamental base for development of optical fluxes propagation possible models in transparent layers of multilayer coatings. Figure 3 shows the widely known zigzag scheme of optical fluxes propagation in the symmetrical optical waveguide (fiber) [8–10] modified introduction of the optical standing wave local interference field areas appearing in reflection positions. The scheme is presented the optical flux propagation corresponding to the multiple consecutive total internal reflection mechanism. For the effective flux propagation, it is a need the consecutive reflection phasing. The geometrical modeling of this demand allowed to receive the characteristic equation, which describes the optical Eigen modes for the symmetrical optical waveguide [16]:

$$\frac{4\pi d}{\lambda_0}\sin\theta_1 + 2\delta = -2m\pi \tag{6}$$

where d is the light guide channel width, θ_1 is the reflection angle corresponding to geometry shown on figure 3, δ is the phase shift at TIR, m is the set of whole numbers defining Eigen modes. The consecutive reflection phasing can be observed for the set of specific angles, which was called as "magic" ones. The propagation effect of optical flux presented on the figure 3 is corresponded to case when the width of light guide channel is large as the transverse size of optical standing wave local interference field areas (d>L/2). The every reflection act excites the interference field in material of the light guide layer external covers. The phasing of reflections does not lead to the covers material reexcitation, and the optical flux propagates through the light guide layer almost without attenuation. When the reflections phasing is broken the great attenuation arises owing to the energy losses appearing.

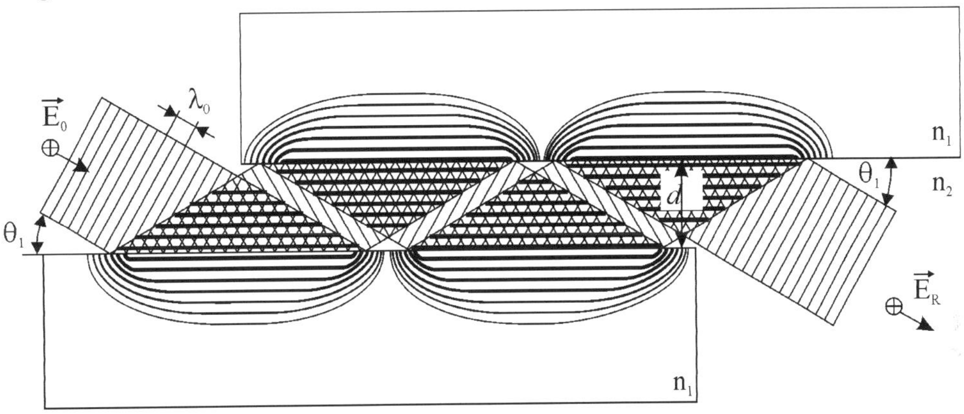

Figure 3. Scheme of the mechanism of multiple consequence total internal reflection displaying the zigzag form in light guide layer of material with enhanced refraction index (n_2>n_1). The Goos-Hanchen shift is not taking account; d is the width of light guide layer.

The optical standing wave period excited in local interference fields is defined by the expression (1). Deflection of the light shaft from "magic" directions on $\Delta\varphi_1$ angle leads to change of the standing wave period on the ΔD_φ value defined by the expression:

$$\left|\Delta D_\varphi\right| = \frac{\lambda_0 \sin\varphi_1}{2n_2\cos^2\varphi_1}\Delta\varphi_1 \tag{7}$$

The radiation wavelength variation on the $\Delta\lambda$ magnitude leads to change of the ΔD_λ standing wave period in according with the expression:

$$\Delta D_\lambda = \frac{\Delta\lambda}{2n_2\cos\varphi_1} \tag{8}$$

For the quasimonochromatic radiation, the $\Delta\lambda$ value is the standard characteristics of its monochromatism degree, and its existence defines the local interference fields spread. So, the effective angular width admitting the phasing of consecutive optical reflections can be calculated on base of ΔD_φ and ΔD_λ equating at use the last parameter in doubled value owing to the φ_1 value can be changed by increasing and reduction on $\Delta\lambda$. In the result, we have:

$$\Delta\varphi_1 = \frac{\Delta\lambda \cos\varphi_1}{2\lambda_0 \sin\varphi_1} \tag{9}$$

For calculation of the multiple consecutive TIR mechanism efficiency, we must take into account the light flux propagation at all "magic" angles ($\Sigma\Delta\varphi_i$) and to compare it with intensity of the optical flux captured by the light guide input aperture. The aperture is defined by the double value of critical total internal reflection for interface between mediums with n_1 and n_2 refraction indexes, which is usually near 90°. Quantity of "magic" angles for different constructions of light guides is not exceeded of ten. The $\Delta\lambda/\lambda_0$ magnitude for any light fluxes is characterized by value near 10^{-4}, but for laser sources this relation is smaller as 10^{-7}. Therefore, the simplest calculation shows that the mechanism of multiple consecutive TIR is characterized by the light flux transportation efficiency, which is not exceeded 0.1%. This fact is not correlated with the efficiency of light guides (fiber), which is observed on practice [10].

At analysis of the multiple consecutive TIR mechanism scheme, presented on figure 3 the question appears about possible change of the light flux propagation mechanism in case of the light guide width decreasing down to size of the local interference field areas transverse value and at its further decreasing. The answer was obtained in the systematical experimental investigations result of the planar extended slit clearance width influence on propagation parameters of quasimonochromatic X-ray fluxes [1, 2]. Executed investigations showed that when the slit clearance width was achieved the transverse size of local interference field areas formed in result of the multiple consequence total external reflection of X-ray quasimonochromatic flux the uniform interference field of radiation standing wave in all space of the slit clearance appeared. Because of conditions are similar for the local interference field areas appearing at total external and total internal reflections one can maintain that in the case of light guide width and the transverse size of the local light standing wave interference field equality the uniform interference filled will appear as according with scheme presented on figure 4.

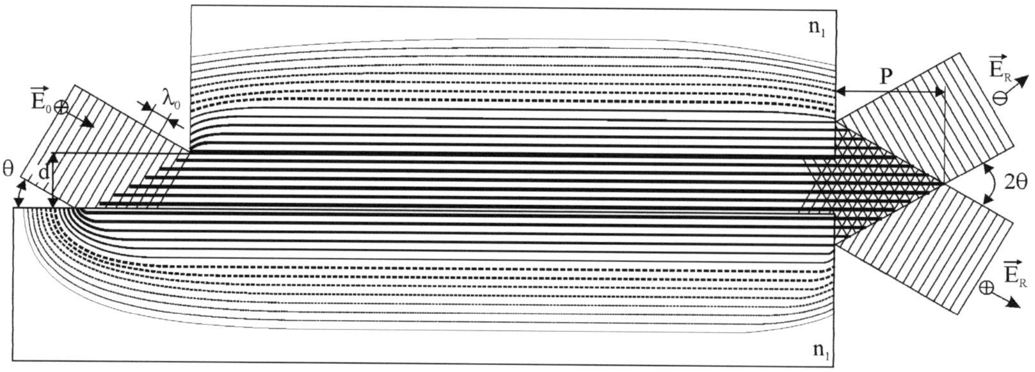

Figure 4. Scheme of the optical standing wave uniform interference field arising in conditions of the light guide layer width is smaller as half of the radiation coherence length propagating through the layer.

The uniform interference field arising in all space of the channel transported the radiation flux is the fundamental feature of radiation fluxes propagation by the waveguide-resonance mechanism [1]. The radiation fluxes transportation by this mechanism has no need of the mode structure and "magic" angles introduction. All optical flux captured by the light guide layer will be transported by the mechanism almost without attenuation owing to the attenuation will be defined by absorption in the transparent light guide material (μ_2) and in materials of covers (μ_1). Light intensity on the light guide output (W(x)) will be defined by the expression:

$$W(x) = W_0 \exp(-x[\alpha\mu_1 + \beta\mu_2]) \tag{10}$$

where W_0 is the initial flux intensity, α and β are the technological parameters, connected with the coating preparation and its are near unit, x is the length of light guide. It is known that the quartz light absorption factor is very small (near 0.0001 cm^{-1}) [18]. Simplest calculation shows that the quartz light guide channel with length near 100 meters will decrease the light intensity on three times. It correlates with practical results [10]. So, one can maintain that the optical waveguides are functioned in frame of the waveguide-resonance mechanism of optical fluxes propagation.

Acknowledgments
The work was carried out in frame of the State task #75-00355-21-00 and at the partial financial support of RFFI (grant #19-07-00271).

References

[1] Egorov E V and Egorov V K 2020 *Radiation fluxes waveguide-resonance propagation phenomenon discovered in result of X-ray nanosize beam formation study* / Chapter 9 in "Electromagnetic propagation and waveguides in photonics and microwave engineering" Ed. by Steglich P London: IntechOpen 145–66
[2] Egorov V and Egorov E 2018 *Optics and spectroscopy* **124(6)** 838–49
[3] Compton A H 1923 *Phil. Mag.* **45** 1121–31
[4] Zachariasen W H 1945 *Theory of X-ray diffraction in crystals* London: Wiley 255 p
[5] Wiener O 1890 *Wied. Ann.* **40** 203–43
[6] Jakiel J and Kantor W 2014 *Eur. Phys. Journal (D)* **68** 305–16
[7] Harrick N J 1967 *Internal reflection spectroscopy* New York: Wiley 324 p
[8] Kapany N S and Burke J J 1972 *Optical waveguides* New York: Academic Press 328 p
[9] Snyder A W and Love J D 1983 *Optical waveguide theory* London: Chapman and Hall Press, 652 p
[10] Kurman S and Deen M J 2014 *Fiber optics communication, fundaments and application* Chichester: Wiley 573 p
[11] Francon M and Slansky S 1965 *Coherence en optique* Paris: Ed. Centre Nat. Rech. Sci. 77 p
[12] Born M and Wolf E 1993 *Principle of optics, electromagnetic theory of propagation, interference and diffraction light* Oxford: Pergamon Press 808 p
[13] Milosevic M 2012 *Internal reflection and ATR spectroscopy* Hoboken, New Jersey: Wiley 239 p
[14] Mandel L and Wolf E 1995 *Optical coherence and quantum optics* Cambridge: Cambridge Univ. Press 631 p
[15] LanterbornW, Kurz T and Wiesenfeldt M 1995 *Coherent optics* Berlin: Springer 294 p
[16] Unger H G 1977 *Planar optical waveguides and fibers* Oxford: Clarindon Press 648 p
[17] Svelto O and Hanna D C 2010 *Principles of lasers* New York: Springer 620 p
[18] Handbook of electronic and photonic materials / 2nd Edition, Ed. by Kasar S and Capper P Leipzig: Springer 2017 1515 p

The applying of the short – term cross correlation in optical diagnostics of melt convection in laser metal deposition

Y N Zavalov and A V Dubrov

Institute on Laser and Information Technologies - Branch of the Federal Scientific Research Centre "Crystallography and Photonics" of Russian academy of Sciences, Svyatoozerskaya 1, 140700, Shatura, Moscow Region, Russia

E-mail: dubrov.av@mail.ru

Abstract. The time dependences of the surface temperature and the surface area of the melt bath during the formation of the single track in the laser metal deposition technology were obtained using a multi-wave optical diagnostics complex. The short - term cross-correlation method was used to analyse the measured parameters of the bath surface, and relationship of the measured parameters with the processes of thermocapillary convection in the melt was shown.

1. Introduction

The technology of manufacturing objects of complex shape using additive laser metal deposition technology (LMD) still has several of disadvantages, in particular, the variability of geometric reproducibility and shape defects [1]. This is due largely to the lower stability of the physical processes of layer-by-layer synthesis compared to traditional production technologies. The effect of laser radiation on the flow of the gas-powder mixture and the surface of the synthesized layer of the object is accompanied by rapid changes in the volume of melt pool and melt surface temperature. The relationship of temporary changes in the geometry of the track formed in the LMD process with changes in the LMD parameters is considered in [2–3]. The importance of studying the convection processes under the action of Marangoni forces is shown in [4] for the technology of selective laser melting. The feature of thermocapillary convection in the LMD technology was under investigation in [5–6]. The experimental study of Benar Marangoni melts convection is indicated to difficult due to the opacity of metals [7]. This paper presents the results of a study of the relationship between the parameters of the melt pool and convection in the melt pool in the LMD technology.

2. Materials and method

The research unit of the additive synthesis was developed at ILIT RAS using the LMD technology. The modification of the part of the unit, the Precitec YC52 laser head, - was carried out at (figure 1) to remove part of the thermal radiation passing through the focusing lens in the opposite direction: between the focusing lens and the collimator of the laser head in the additional housing at the angle of 45° to the optical axis, the plane-parallel optical plate and the plate for compensating for the displacement of the optical axis are located.

The flow of PR-X18N9 powder with the mass flow rate of 8.4 g/min and the granulometric composition (40 ... 100) microns was concentrated through the conical nozzle on the substrate made

Content from this work may be used under the terms of the Creative Commons Attribution 3.0 licence. Any further distribution of this work must maintain attribution to the author(s) and the title of the work, journal citation and DOI.
Published under licence by IOP Publishing Ltd

of mild steel with the thickness of 4 mm. A fibre ytterbium laser LK-400-V was used, the radiation of which passes through the coaxial flow of the gas-powder mixture and focuses near the substrate. Each track with the length of 30 mm was formed at constant values of the distance from the nozzle to the substrate and the position of the beam focus, scanning speed V=8 mm/s, laser power P=240 W. Experiments on the formation of tracks were carried out in the online monitoring mode of the melt pool (MP) using the multi-wave optical diagnostics complex (MODC) installed according to the coaxial scheme.

For spectral selection of laser radiation reflected from the melt surface, as well as for the illumination of plate surfaces at the laser wavelength, multilayer interference coatings of plates are used. In addition, a set of optical filters is used to limit the sensitivity of the video matrix of the video camera to the range of the visible spectrum. The infrared part of the thermal radiation of the heated surface of the melt pool is focused by the lens on a multi-channel sensor (figure 1b) of the pyrometer of the spectral ratio. The spectral sensitivity windows of the K12729-010K photosensor (Hamamatsu Photonics), taking into account the additional optical filters used, lie in the ranges (1.74...2.1) microns for the first type of photodiode and (1.15...1.46) microns for the second type of photodiode. Commercially available pyrometers are designed to measure the temperature of bodies by its thermal radiation within the so – called Field of view (FOV). The field size is greater than 3 mm in the typical case, and 0.5 mm in the special version [8]. If the size of the melt pool is less than FOV, the pyrometer readings decrease, in addition, the unevenness of the temperature distribution within the FOV also leads to the methodological error when measuring the temperature. In the multichannel pyrometer we used, the FOV of each channel of the pyrometer used was 120 microns, taking into account the increase in the optical system, the use of six channels allows us to simultaneously measure the temperature at the site of about 1.5 mm2. The distance between the optical fibers of 0.35 mm corresponds to the distance on the surface of the BP of about 0.54 mm.

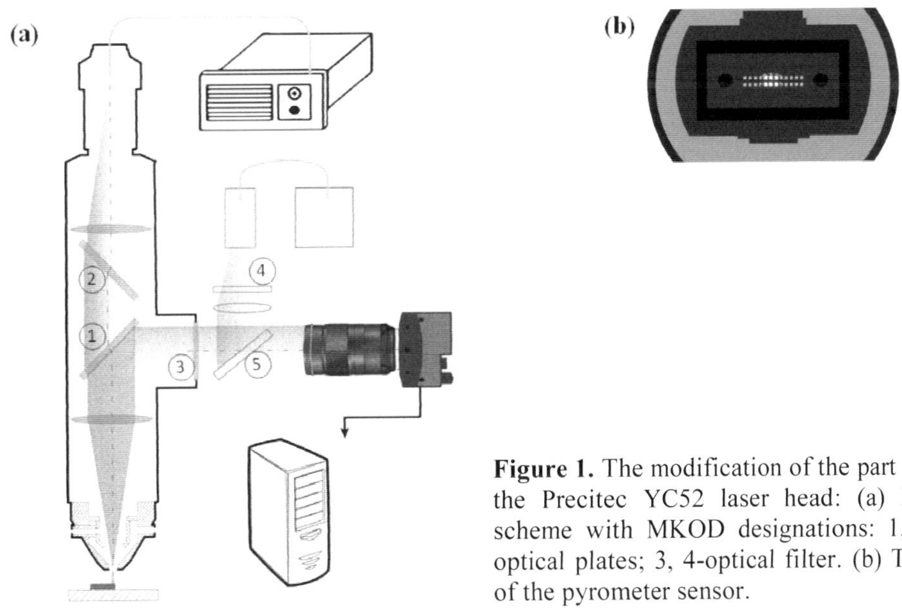

Figure 1. The modification of the part of the unit, the Precitec YC52 laser head: (a) Installation scheme with MKOD designations: 1, 2 and 5-optical plates; 3, 4-optical filter. (b) The scheme of the pyrometer sensor.

The MODC allows obtaining data on the geometry of the MP surface, as well as the maximum temperature on the surface T during the formation of the track under conditions of reflected laser radiation, luminescent glow of metal oxides or salts on the surface of the melt pool, as well as ionized metal vapors. An overview of the methods of optical diagnostics is given in [9]. To determine the geometric characteristics of MP in the MODC, a high-speed camera Mikrotron 3110 was used [10]. The total magnification of the "focal lens – tele lens x2.2 – lens" system was x1. 4. Images of the MP surface with the spatial resolution of about 10 microns, the frame rate of up to 3000 Hz with the frame size of 560 x 280 pixels were obtained. In real time, the following geometric characteristics of MP are distinguished: the area and perimeter of the MP region; the lengths of the small and large axes of the equivalent ellipse. To digitize the photocurrents of the set of diodes when determining the maximum temperature T_{max} on the melt pool surface, two universal LTR11 ADC modules are used as part of the LTR-EU-8 crate (L-CARD). The data digitization depth is 14 bits, the sampling frequency is 5 kHz.

Figure 2 shows the track profile, as well as oscillograms of changes in the area of the melt pool A and temperature (2500 K-T[K]) (dotted line). The latter expression decreases with the growth of T. As follows from figure 2 A and T in some areas (highlighted for example) are highly anticorrelated, the temperature increase is accompanied by the drop in the melt area. The ordinate in figure 2 is calculated as $x= V*t$, t is the current time associated with the storyboard of the video sequence, which allows you to link the video sequence and the track profile. The average track height was 0.41 mm, the standard deviation was 12%.

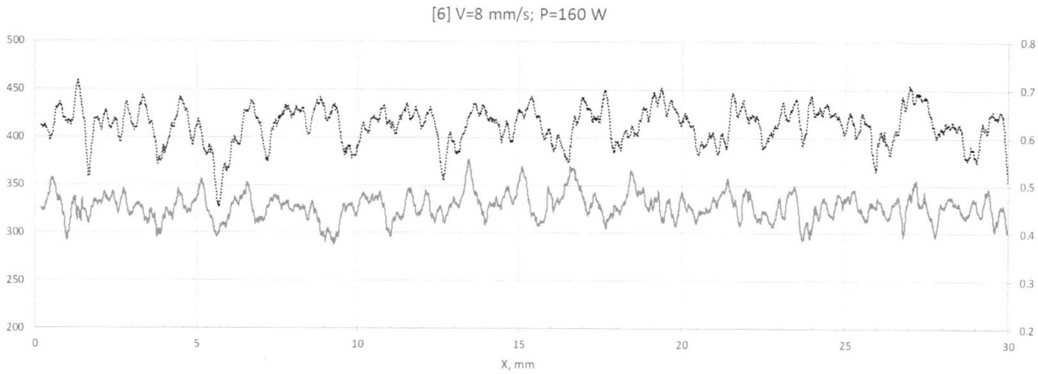

Figure 2. The oscillograms of changes in the area of the temperature (2500 K-T[K]), dotted line, and melt pool A (right scale) as well as the track profile.

To study the correlation of changes in the area of the melt pool A and temperature, the method of cross – correlation analysis of short-term cross correlation (STCC) samples was used [11]. It makes sense to find correlations only for a small sample in the long-time video sequence of data. The length of the melt pool is less than 2D [10] and the short term defines as $\pm D/V$, where the width of the melt pool D is 0.76 mm. The STCC T–A data is presented in figure 3 with the averaging of the initial data of 3 ms and the short sample length of 30 samples. Blue is positive correlation values, yellow is negative. In figure 2, the scale along the ordinate axis is ± (3 ms·30 samples ·V) = ±0.72 mm. There are areas of anticorrelation <T–A>, highlighted by the oval in Fig.3. Laser radiation in the LMD process heats a thin layer of the metal surface, and the melt convection occurs under the action of Marangoni forces when the surface of the melt pool is inhomogeneously heated by laser radiation. The surface temperature is determined by the convection efficiency. As its efficiency increases, the temperature on the surface decreases, at the same time the phase boundary shifts towards an increase in the contact area of the melt with the non-molten surface of the substrate with an increase in the heat flow through the phase boundary, taking into account the high thermal conductivity of the metal. Thus, the decrease in temperature should be accompanied by the increase in the volume of the melt pool, including A. Conversely, with the decrease in the efficiency of convection, the temperature on the surface increases, the heat flow into the interior decreases, the boundary with the solid phase crystallizes and shifts, the melt area decreases.

In other areas, there is no correlation, convective processes are too weak, or they did not have time to develop, for example, near the ordinate axis at the initial section of the track formation (the abscissa axis is highlighted in pink in this case). In this case, the convection efficiency is low, and therefore, with the constant flow of powder to the surface of the melt pool, the height of the formed track increases. In the case of the sharp increase in the efficiency of heat and mass transfer, dips in the track height profile are observed.

We also see sections of equidistant bands parallel to the abscissa (highlighted) in figure 3. These are fluctuations in the measured parameters. The feature of the correlation analysis, it allows you to determine the fluctuations of the parameters against the background of interference [11]. The oscillation frequencies vary in the range of (80...140) Hz. The duration of the oscillatory process is slightly higher than the time to travel the distance of the size of the laser beam on the substrate, which is $2\omega \approx 1.3\ mm$ at the $1/e^2$ level [3]. The correlation of changes in the area of the melt pool A and temperature is associated with the convection process, the fluctuations of these parameters reflect the realized structure of vortex motion during thermocapillary convection [12], which allows us to evaluate the efficiency of heat and mass transfer.

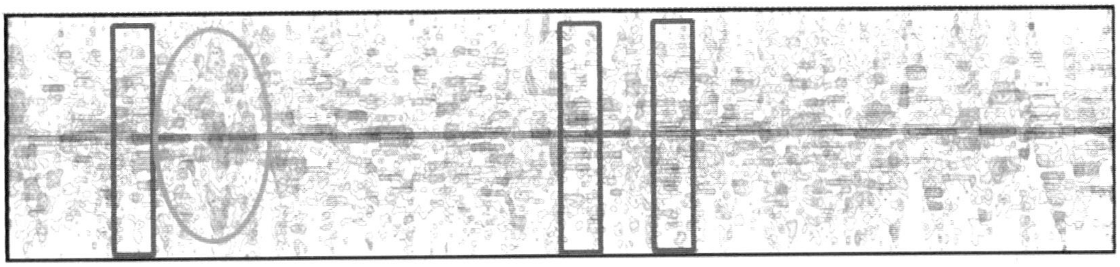

Figure 3. STCC <*T–A*>. Blue is positive correlation values, yellow is negative.

3. Results and discussion

Let us imagine the structure of convection developed under the action of Marangoni forces in the form of two vortices, twisted oppositely with the axis of rotation parallel to the scanning direction. The vortex length L= 11 mm is slightly less than length of the molten pool, the rotation frequency is $2\pi\times F$, where $F\approx 100$ Hz is the frequency of periodic changes T and A observed in figure 3. In this case, the maximum speed of the melt movement is U:

$$U = 2\pi \cdot F * R, \tag{1}$$

where R $\approx \omega$ - the vortex radius.

Let us estimate the size of the melt a in the cross-section of the vortex, such that inside the square side the melt will have time to cool down in the lower part of the vortex or heat up in its upper part:

$$a \approx \sqrt{\frac{\chi}{F}} \approx 0.2\ mm, \tag{2}$$

where $\chi \approx 4\frac{mm^2}{s}$ thermal conductivity.

The value of a is approximately 2/3R, so the melt will have time to give off heat or heat up, fulfilling the condition of convective heat exchange with the phase interface surface [13].

4. Conclusions

Thus, the multi-wave optical diagnostics complex at the laser metal deposition unit was used to determine the temporary changes in the temperature and geometric parameters of the melt pool during the formation of the track made of stainless steel powder PR-X18N9. The spatial - temporal distribution of the correlation of temperature changes on the melt surface with the simultaneous change in the area of the melt pool occurring at the distance from the heating region is obtained using the method of cross-correlation analysis of short samples. It is shown that at certain sections of the track formation, the measured parameters were anticorrelated or experienced fluctuations with frequencies in the range of (80…140) Hz. The relationship of the parameters of the melt pool surface with the processes of thermocapillary convection of the melt is shown, and the estimate of the melt velocity during convection under the action of Marangoni forces is obtained. The estimation of the efficiency of convective cooling of the melt is in good agreement with the data of temperature measurement on the melt surface.

Acknowledgements
The work was supported by the Ministry of Science and Higher Education of the Russian Federation within the framework of the State Task of the Federal State Budgetary Institution "Crystallography and Photonics" of the Russian Academy of Sciences in terms of setting the problem, as well as with the grant support of the Russian Foundation for Basic Research in terms of the developed methods of optical diagnostics, grant No. 18-29-03249, as well as in terms of conducting experiments and the experimental results (grant No. 20-21-00158).

References

[1] DebRoy T, Wei H L, Zuback J S, Mukherjee T, Elmer J W, Milewski J O, ... and Zhang W 2018 *Progress in Mater. Sci.* **92** 112–224
[2] Sammons P M, Bristow D A and Landers R G 2013 *J. of Manuf. Sci. and Eng.* **135(5)** 054501
[3] Zavalov Y N, Dubrov A V, Ivanov I M and Dubrov V D 2019 *J. Phys. Conf. Ser.* **1396** 012046
[4] Rombouts M M, Kruth J–P, Froyen L and Mercelis P 2006 *CIRP annals* **55** 187–92
[5] Lee Y S and Farson D F 2016 *Int. J. Adv. Manuf. Technol.* **85** 1035–44
[6] Zheng B, Haley J C, Yang N, Yee J, Terrassa K W, Zhou Y, ... and Schoenung J M 2019 *Mater. Sci. Eng.* **764** 138243
[7] Boeck T and Vitanov N K 2002 *Phys Rev E* **65** 037203
[8] Kienitz U 2014 *J. of Sens & Sens. Syst.* **3(2)** 265–71
[9] Sun Z, Guo W and Li L 2020 *Opt. & Laser Technol.* **129** 106280
[10] Zavalov Y N, Dubrov A V and Makarova E S 2020 *J. Phys. Conf. Ser.* (IOP Publishing) **1713** 012048
[11] Chatfield C 2016 *The Analysis of Time Series: An Introduction* (CRC Press)
[12] Getling A V 1998 *Structures and dynamics Advanced series in nonlinear dynamics* (Singapore: World Scientific) vol 11
[13] Yang L X, Peng X F and Wang B X 2001 *Int. J. of Heat and Mass Transfer* ***44*(23)** 4465–73

Bioactive coatings on 3D printed titanium implants with a complex internal structure for bone replacement

A I Kozelskaya[1], S Rutkowski[1,2], A S Gogolev[1], S G Chistyakov[1], I B Krasovsky[3], A A Zheravin[4] and S I Tverdokhlebov[1]

[1] Tomsk Polytechnic University, 634050, Tomsk, Russia
[2] Tomsk Scientific Center of the Siberian Branch of the Russian Academy of Sciences, 10/4, Akademicheskii Prospekt, Tomsk, 634055, Russia
[3] Limited Liability Company «Logeek[s]MS», 630055, Novosibirsk, Russia
[4] E. Meshalkin National Medical Research Center of Ministry of Health of Russian Federation, 630055, Novosibirsk, Russia

E-mail: kozelskayaai@tpu.ru

Abstract. The micro-arc oxidation method was applied to modify the surface of 3D printed titanium implants with a complex internal structure. Two different electrolyte solutions were used for the surface modification, for which the respective working parameters of the micro-arc oxidation process were developed. Surface coatings formed with these parameters on the 3D implants have the same chemical composition and have the same surface morphology as surface coatings on 2D substrates. The measurement of the coating thickness using the X-ray microtomography demonstrates that this method is a useful tool for the thickness control of porous surface coatings at the inside and outside of the 3D titanium implants.

1. Introduction

Additive technologies, also known as additive manufacturing methods, are widely used in different industries. Especially the 3D printing process is known as one of these technologies, as well as the laser beam melting process, the layer-by-layer electron beam melting process and printing by lamination in which individual 3D structures are producible [1]. The 3D printing process is becoming more and more important, especially for medical applications. Using magnetic resonance and X-ray tomography, medical data can be obtained for the fabrication of individual 3D implants [2].

The additive technologies enable the production of individual and precisely fitting implants. Due to the additional possibility of adapting the inside of the implant with different filling patterns and thus the porosity, the integration with the bone tissue is also improved by the inside of the implant. Therefore, the development of methods and technologies for the modification and quality control of highly porous 3D tissue-engineered structures is one of the priority directions in material science and applied interdisciplinary science for the introduction of advanced regenerative technologies into clinical practice.

The formation of bioactive coatings on the surface of medical implants leads to an increase in the speed and quality of bone integration with the surgically installed implant. In addition, implants with calcium phosphate containing coatings are already being used successfully in medical practice to replace bone defects. However, the formation of such coatings on the surface of implants with a

Content from this work may be used under the terms of the Creative Commons Attribution 3.0 licence. Any further distribution of this work must maintain attribution to the author(s) and the title of the work, journal citation and DOI.
Published under licence by IOP Publishing Ltd

complex shape and with a high porosity requires special methods. One of these methods for coating complex-shaped implants is the micro-arc oxidation. The aim of this work was to develop working parameters (hereinafter called as working modes) for the modification of highly porous 3D titanium implants by micro-arc oxidation, as well as to select and test a method for determining the thickness and uniformity of the coating over the entire surface of the samples.

2. Materials and methods

The 3D printed titanium test samples (fabricated by the company «LOGEEK^S MS», LLC, Novosibirsk, Russia) were made of a titanium alloy and have the following dimension: high: 15 mm, diameter: 10 mm. All samples were subjected to chemical etching prior to the formation of the coatings. The coatings were formed in two different electrolytes. For the first electrolyte, the aqueous solution consists of calcium acetate ($Ca(CH_3COO)_2 \cdot H_2O$, c = 15.6 g/L, purchased from Component-Reaktiv, LLC, Moscow, Russia) and sodium dihydrogen phosphate dihydrate (Na_2HPO_4, c = 36.8 g/L, purchased from VEKTON, JSC, St. Petersburg, Russia) with the addition of phosphoric acid (H_3PO_4, purchased from Component-Reaktiv, LLC, Moscow, Russia) at a concentration of 2.5 ml/L. A supersaturated calcium oxide (CaO, purchased from Component-Reaktiv, LLC, Moscow, Russia) solution in a 10% aqueous solution of phosphoric acid (H_3PO_4) with the addition of a dispersed phase of hydroxyapatite (purchased from Fluidinova, Maia, Portugal) in an amount of up to 10 g/L was used as the second electrolyte. Hereafter, 'MAO_1' refers to the 3D titanium samples with the coating formed in the first electrolyte and 'MAO_2' refers to the 3D titanium samples with the coating formed in the second electrolyte. The coating formations were carried out on the setup "Micro-arc oxidation complex" designed in the Laboratory for Plasma Hybrid Systems, The Weinberg Research Center, School of Nuclear Science & Engineering of the Tomsk Polytechnic University (Tomsk, Russia). For the surface modification of the 3D titanium implants with the micro-arc oxidation process, a separate working mode was developed for each of the two electrolyte solutions. The working mode for the MAO_1 electrolyte solution was as follows: voltage – 350 V, voltage rise rate – 2 V/s, pulse repetition rate – 220 Hz, pulse duration 140 μs, coating formation time – 15 min. For the MAO_1 electrolyte solution, the working mode parameters are as follows: voltage – 320 V, voltage rise rate – 3 V/s, pulse repetition rate – 230 Hz, pulse duration –120 μs, coating formation time – 50 min. The temperature of electrolyte solution in both cases did not exceed 15 °C during the micro-arc oxidation process.

The surface morphology (as microscope images) and the elemental composition (as energy-dispersive X-ray spectroscopy, EDS) of the coatings were analyzed using a Quanta 200 3D scanning electron microscope (FEI Company, Hillsboro, Oregon, USA) equipped with an energy-dispersive X-ray spectrometer (JSM 5900LV, JEOL Ltd., Tokyo, Japan).

Measurements of the thickness and uniformity of the coating deposition on the 3D titanium implant samples were carried out using an X-ray microtomography scanner (micro-CT), which was designed and built-up in the international research laboratory for X-Ray optics at the Tomsk Polytechnic University (Tomsk, Russia). Micro-CT micrographs of the coated samples were received by a self-build X-ray scanner utilized with a Hamamatsu microfocus X-ray source (L9181-02, Hamamatsu Photonics, Hamamatsu City, Japan) with a tube voltage of 130 kV and a current 80 μA without beam filtration [3]. A flat panel detector (Mark2430C, PRODIS.NDT, pos. Malakhovka, Ovrazhki, Moscow Oblast, Russia) was used for image acquisition [4]. The voxel (grid value in a 3D room) size is 3.21 μm, taking into account the geometric magnification of 15.42. An image acquisition time of 900 ms was set, resulting in a total scan duration of approximately 15 minutes for 1000 projections. For the reconstruction of the tomograms was performed using a standard back projections algorithm with a flat field, ring artifacts and a beam hardening correction. To analyze of 3D tomograms and to plot of the thickness distribution by size, The VGSTUDIO MAX 3.4.5 software (Volume Graphics, Heidelberg, Germany) utilized with the module "Wall Thickness Analysis" [5].

3. Results and discussion

Figure 1 shows the macroscopic appearance as a photograph of the 3D titanium samples with the calcium phosphate coatings, as well as SEM micrographs of their microscopic surface morphology.

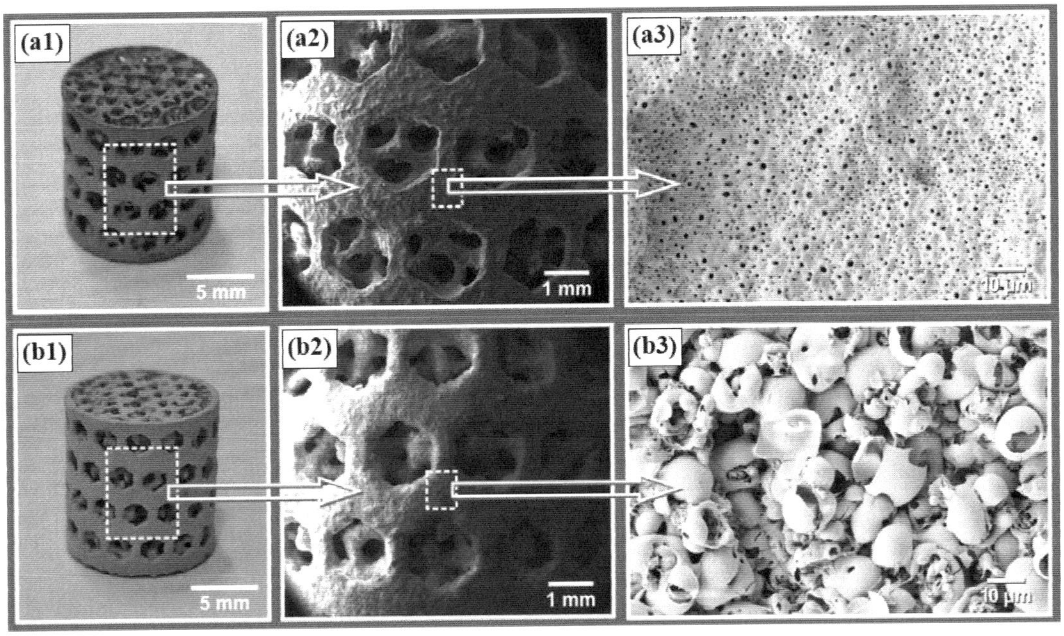

Figure 1. The macroscopic appearance of the 3D titanium implant samples, shown by photographs with the MAO_1 coating **(a1)** and the MAO_2 coating **(b1)**, and their corresponding microscopic surface appearance shown by SEM images in different magnifications **(a2, b2** and **a3, b3)**.

Surface coatings formed in different electrolyte solutions differ in color and structure. Coatings formed with the MAO_1 are brown and coatings formed with the MAO_2 solution are gray (figure a1 and b1). As figures a2 and b2 show, the coatings cover the entire sample evenly.

Coatings produced by the micro-arc oxidation process usually have a porous structure. However, the composition of the electrolyte solution and the micro-arc working modes affect the morphology of the fabricated coatings. The surface morphology of the MAO_1 coated 3D implant sample consists of crater-like formations with a pore in the center of each crater (figure a3). On the surface of the MAO_1 coating, the number of pores in the surface area of 1000 μm^2 is 172.5 ± 12.6 and the average diameter of these pores is 0.82 ± 0.20.

The surface of the 3D implant sample coated with the MAO_2 solution has a rougher structure, which is caused by hollow spherical particles and hemispheres with bigger pores (figure b3). The number of the spherical particles that are situated in the surface area of 1000 μm^2 is 19.6 ± 2.4 with an average size of (10.71 ± 2.50) µm. In this area are pores with a number of 25.6 ± 2.3 and an average diameter of (3.76 ± 1.08) µm.

Table 1 shows the elemental composition of the coatings, which corresponds to the composition of the utilized electrolyte solutions and consists of the elements carbon (C), oxygen (O), phosphorus (P) and calcium (Ca). Aluminum (Al) and titanium (Ti) are present in the elemental composition measurements of the coatings due to the substrate consisting of titanium alloy.

The Ca/P ratio in the applied surface coatings is 0.89 ± 0.05 for the MAO_1 coating and 0.37 ± 0.04 for the MAO_2 coating. These determined Ca/P ratios are in accordance with coatings formed on flat substrates, which were reported in references [6]. These results prove the finding of the optimal micro-arc oxidation working modes, whereby the physico-chemical properties of the coatings on the 3D titanium implant samples are comparable with known coatings on flat structures.

Table 1. Elemental composition of the formed coatings obtained by EDS, at/%.

Sample	C	O	Al	P	Ca	Ti
MAO_1	4.79 ± 1.01	50.58 ± 1.59	1.00 ± 0.14	1.89 ± 0.23	1.69 ± 0.15	40.07 ± 2.85
MAO_2	19.62 ± 0.89	46.34 ± 1.94	0.43 ± 0.14	16.25 ± 0.28	6.02 ± 0.80	11.35 ± 1.01

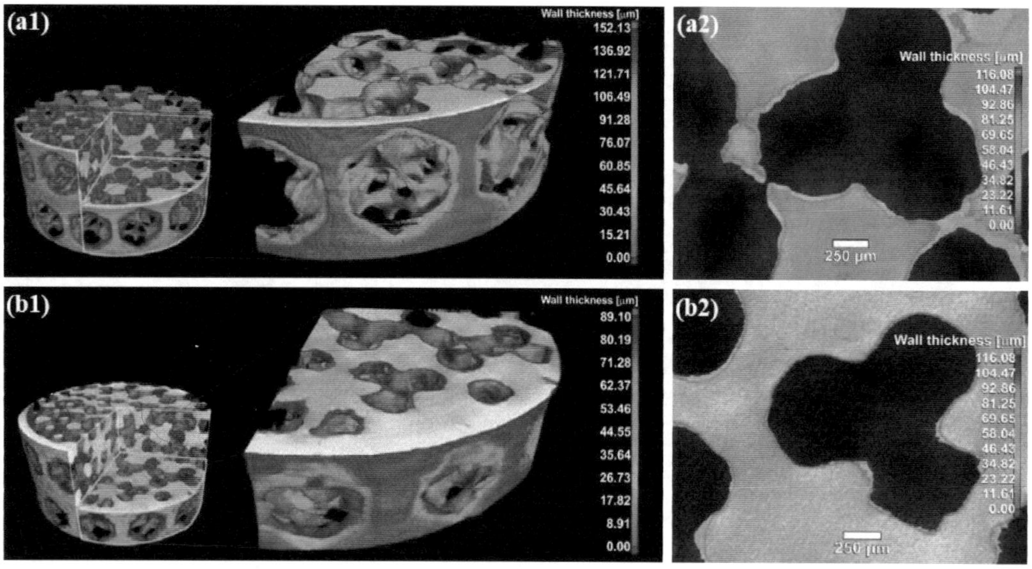

Figure 2. X-ray microtomograms of the scanned implant sections of the calcium phosphate containing surface coatings MAO_1 (**a1-a2**) and MAO_2 (**b1-b2**).

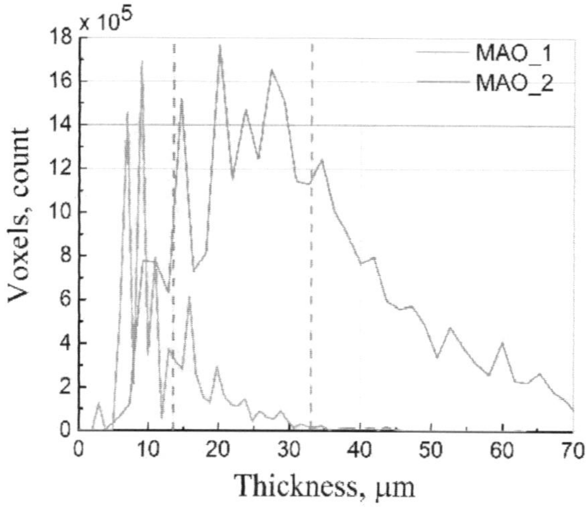

Figure 3. Layer thickness distribution in the region of interest with a size of 2.5 × 2.5 × 1.5 mm for the surface coatings MAO_1 and MAO_2 on the 3D titanium implant samples.

The thickness and heterogeneity of the coatings on the 3D titanium implant samples were determined by X-ray microtomography (figure 2). This method provides information about the internal structure of the highly porous and complex-shaped 3D samples with calcium phosphate coatings and visualizes them with high-definition images. Figure 2 shows that both the inside and outside of the 3D titanium implant samples are covered with a calcium phosphate coating. For the MAO_1 coating, the thickness is in the range from 0.12 μm to 116.08 μm and for the MAO_2 coating, the thickness is in the range from 0.12 μm to 63.37 μm. The lowest value for the measured coating thickness represents the lower sensitivity limit of the utilized X-ray microtomography device.

Figure 3 summarizes the results of the X-ray microtomography as thickness distribution diagram for the surface coatings MAO_1 and MAO_2 on the 3D titanium implant samples in the region of interest with the size of 2.5 × 2.5 × 1.5 mm over the count of voxels with a size of 3.21 μm. As the result, the average thickness of the MAO_1 coatings is (33.01 ± 15.75) μm, while for the MAO_2 coatings the average thickness is (13.39 ± 7.17) μm. Figure 2 and 3 demonstrate that the MAO_1 coatings have a higher layer thickness distribution and less heterogeneity compared to the MAO_2 coatings.

4. Conclusions

As a first result, the required parameters of the micro-arc oxidation for the modification of 3D printed titanium implants were determined. The applied working modes for two different electrolyte solutions to form calcium phosphate containing surface coatings make it possible to obtain coatings that are chemically similar to those on flat titanium implant samples. The results of the elemental composition analysis and the obtained Ca/P ratios are in agreement with the actual literature in this research field. Layer thicknesses and the heterogeneity of the formed surface coatings on the 3D titanium implant samples with a complex internal structure were successfully investigated applying the X-ray microtomography method. The MAO_1 coatings have a higher layer thickness distribution and less heterogeneity compared to the MAO_2 coatings. Coatings formed via the MAO_1 electrolyte solution having a thickness in the range of 0.12 μm to 116.08 μm and average thickness of (33.01 ± 15.75) μm. The thickness of the coatings formed via the MAO_2 electrolyte solution is in the range from 0.12 μm to 63.37 μm with an average thickness of (13.39 ± 7.17) μm.

Acknowledgments

This research was carried out within the state assignment of the Ministry of Health of the Russian Federation (theme No 121032300337-5).

References

[1] Gokuldoss P K, Kolla S and Eckert J 2017 *Materials* **10** 672
[2] Javaid M and Haleem A 2018 *J. Clin. Orthop. Trauma* **9** 202–6
[3] Microfocus X-ray source L9181-02 URL - https://www.hamamatsu.com/eu/en/product/type/L9181-02/index.html]
[4] Dynamic high resolution x-ray flat panels Prodis. Mark (CMOS) - PRODIS. NDT URL - https://prodis-tech.ru/en/dynamic-high-resolution-x-ray-flat-panels-prodis-mark/
[5] VGSTUDIO MAX URL - https://www.volumegraphics.com/en/products/vgsm.html
[6] Kozelskaya A, Dubinenko G, Vorobyev A, Fedotkin A, Korotchenko N, Gigilev A, Shesterikov E, Zhukov Y and Tverdokhlebov S 2020 *Coatings* **10** 1113

General principles for describing electronic and proton multiple scattering processes in solids

V P Afanas'ev, L G Lobanova, D N Selyakov and M A Semenov-Shefov

National Research University «Moscow Power Engineering Institute», 111250, Moscow, Russia

E-mail: v.af@mail.ru

Abstract. Analytical solution for the reflected light ions Pass Length Distribution Function (PLDF) equation is obtained. Reflected ions energy spectra calculated on the basis of the developed method shows satisfactory agreement with experimental data. The effectiveness of the developed methodology in the procedure for verifying the stopping power value is indicated.

1. Introduction

Consistent and accurate description of the interaction processes of accelerated ion fluxes with the solid surfaces is necessary both for the implementation of technologies for modifying and doping materials, and for signals descriptions from a wide range of ion-scattering spectroscopy techniques.

The most active research in this area was carried out in the 80s of the last century [1-5], which first was associated with the problem of the "first wall" in controlled thermonuclear fusion installations. If we look at the analytical models used to describe the process of ion scattering in solids [1-3], then, due to their extreme simplification, the question arises about the possibility of their use for a qualitative description of the processes of reflection of ions with energies from units to tens of keV.

There are two main characteristics of the scattering of light ions, which are commensurate in the indicated energy range: the transport path l_{tr} and the residual path $R_0 = \frac{T_0}{\bar{\varepsilon}}$ (T_0 is the initial ion energy, $\bar{\varepsilon}$ - stopping power):

$$R \sim l_{tr}. \tag{1}$$

Formula (1) indicates that the use of Strait Line Approximation (SLA), which is traditional for ions energies from 1 to tens keV, is unacceptable. The situation is similar to which is typical for electrons [6-8]. Papers [6, 8] describe the most effective technique for describing electronic spectra.

In this work, we adapt the method presented in [6, 8] to the problems of light ions scattering of in solids.

2. Methods of Reflected Light ions energy spectra calculation

Let us consider the problem of electrons and light ions elastic scattering in solids. Solving the boundary value problem for the elastic transfer equation by the invariant immersion method leads us in azimuthally symmetric situations to the formula:

$$\left(\frac{1}{\mu}+\frac{1}{\mu_0}\right)\left(1+\frac{\partial}{\partial\tau}\right)S(\tau,\mu,\mu_0) = \tag{2}$$

Content from this work may be used under the terms of the Creative Commons Attribution 3.0 licence. Any further distribution of this work must maintain attribution to the author(s) and the title of the work, journal citation and DOI.
Published under licence by IOP Publishing Ltd

$$x(\mu,\mu_0)+\int_0^1 x(\mu,\mu')S(\tau,\mu',\mu_0)\frac{d\mu'}{\mu'}+\int_{-1}^0 S(\tau,\mu,\mu')\,x(\mu',\mu_0)\frac{d\mu'}{\mu'}+$$
$$\int_0^1\int_{-1}^0 S(\tau,\mu,\mu')x(\mu',\mu'')S(\tau,\mu'',\mu_0)\frac{d\mu'}{\mu'}\frac{d\mu''}{\mu''},$$

where $S(\tau,\mu,\mu_0)$ is the reflection function determined by (a): the scattering geometry: μ,μ_0 are the cosines of the viewing and probing angles; (b): the dimensionless path length of the ion in the target $\tau=\frac{z}{l_{el}}$, l_{el} is the mean free path between the elastic collisions, $x(\mu,\mu_0)$ is normalized differential elastic scattering cross section.

In the case of conservative scattering, when the channel of inelastic scattering is excluded, it is possible to obtain an analytical solution of equation (2), the solution procedure, justification of the assumptions made based on comparison with the exact numerical solution (2) are presented in [8]. The solution obtained is absolutely the same for both electrons and light ions!

But in the case of light ions, due to the fact that the screening parameter of the process of light ions elementary scattering is orders smaller than electrons, the assumptions made to obtain an analytical solution work in a wider energy range. The Pass Length Distribution Function, the distribution of reflected atomic particles over the path lengths in the target:

$$S(\tau,\mu,\mu_0)=\frac{|\mu|\mu_0}{|\mu|+\mu_0}\frac{1}{\tau}\sum_{l=0}^{\infty}\frac{2l+1}{2}P_l\left(\mu\mu_0+\sqrt{1-\mu^2}\sqrt{1-\mu_0^2}\right)\left(e^{-(1-x^l)\tau}-e^{-\tau}\right), \quad (3)$$

where x^l are the coefficients of the series expansion of the normalized differential cross section of elastic scattering in a series in Legendre polynomials. The sections are constructed on the basis of the potential of krypton carbon - KrC [1-3].

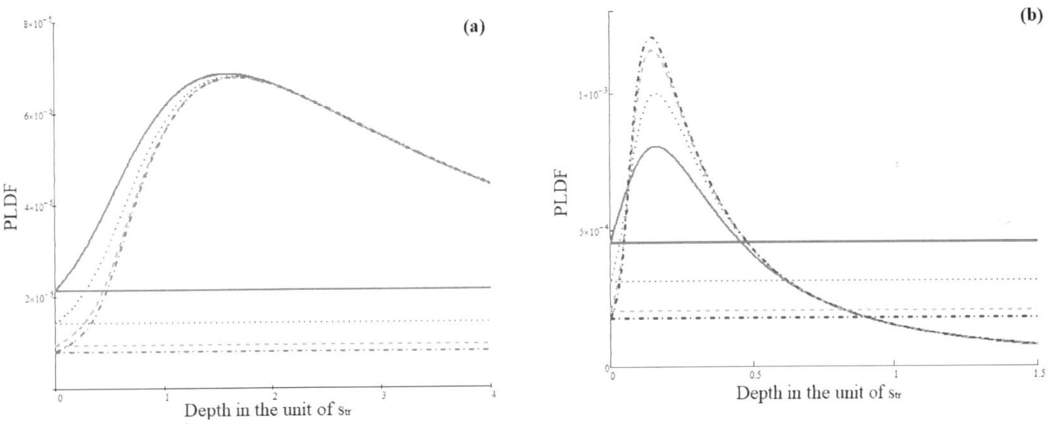

Figure 1. Part **a**: PLDF for $\eta=10^{-3}$ – red solid line, $\eta=5\cdot10^{-5}$ – blue dotted line, $\eta=5\cdot10^{-6}$ – green dashed line, $\eta=5\cdot10^{-7}$ – purple dashed line. Normal sounding, scattering angle 135^0. Part **b**: PLDF for $\eta=10^{-3}$ – red solid line, $\eta=5\cdot10^{-5}$ – blue dotted line, $\eta=5\cdot10^{-6}$ – green dashed line, $\eta=5\cdot10^{-7}$ – brown dashed line. Incidence angle 20°, scattering angle 110^0. Rutherford scattering cross section. Horizontal lines coming from PLDF plots obtained on the basis of (3), at point $s=0$, represent the distributions over distances in the Straight Line Approximation ($PLDF_{SLA}$ dependence).

The area of zero paths (figure 1) describes protons reflected from the target surface. Setting $s=0$ in (3), we obtain:

$$S(0,\mu,\mu_0)=\frac{|\mu|\mu_0}{|\mu|+\mu_0}x_{el}\left(\mu\mu_0+\sqrt{1-\mu^2}\sqrt{1-\mu_0^2}\right). \quad (4)$$

Formula (4) is of decisive importance in ion backscattering spectroscopy. It describes PLDF in the Strait Line Approximation (SLA):

$$S_{SLA}(\tau,\mu,\mu_0)=S(0,\mu,\mu_0). \quad (5)$$

An important result following from calculations based on (3) is the transition to a constant universal form of the PLDF distribution as the screening parameter tends to zero. Moreover, the specified universal form is determined only by the geometry of the experiment! In the case shown in figure 1, the PLDF maximum is observed in the region of the order of one and a half transport paths.

The differences in PLDF obtained on the basis of successive accounting for multiple elastic scattering of ions and in their neglect (SLA approach) is illustrated in figure 1. The possibility of using the SLA approach opens up in cases when the stopping path of a particle is significantly less than its transport path:

$$R_0 \ll l_{tr}. \tag{6}$$

Attempts to use (3) in the case of lead to absurd results, for which the experimental and calculated in SLA spectra are presented on a logarithmic scale [9].

We will determine the energy spectra of reflected particles in accordance with the formula:

$$S(\Delta, \mu, \mu_0) = \int_0^{\infty} dz \cdot S\left(\frac{z}{l_{el}}, \mu, \mu_0\right) T\left(\frac{z}{l_{in}}, \Delta\right). \tag{7}$$

$T(\tau,\Delta)$ in the case of ion scattering is determined only by two moments of the cross section of inelastic scattering $\bar{\varepsilon}$ and $\overline{\varepsilon^2}$ – the average energy losses per length unit and the mean square of energy losses or straggling, which are dictated by the laws of conservation of energy and momentum, limiting the energies transferred from the ion to the electronic subsystem:

$$T(z,\Delta) = \frac{1}{\sqrt{2\pi\overline{\varepsilon^2}z}} e^{\left[-\frac{(\Delta-\bar{\varepsilon}z)^2}{2\overline{\varepsilon^2}z}\right]}. \tag{8}$$

Substituting PLDF (3) into the formula for calculating the energy spectra of reflected ions (7), we obtain, in the approximation of continuous deceleration, when $T(z,\Delta) = \delta(\Delta - \bar{\varepsilon}z)$:

$$S(\Delta, \mu, \mu_0) =$$

$$\frac{|\mu|\mu_0}{|\mu|+\mu_0} \frac{(1-x_1)(1-\Delta)}{\Delta\bar{\varepsilon}} e^{\left(-\frac{A}{\sqrt{1-\Delta}}\right)} \sum_l \frac{2l+1}{2} P_l\left(\mu\mu_0 + \sqrt{1-\mu^2}\sqrt{1-\mu_0^{\,2}}\right) \left[e^{-\frac{\Delta(1-x_l)\sigma^*}{(1-x_1)(1-\Delta)^{5/2}}} - e^{-\frac{\Delta\sigma^*}{(1-x_1)(1-\Delta)^{5/2}}}\right]. \tag{9}$$

The dimensionless parameter σ^* appeared in formula (9), which is decisive for the analytical description of the process of reflected ions energy spectra formation:

$$\sigma^* = \frac{E_0}{l_{tr} \cdot \bar{\varepsilon}}, \tag{10}$$

where $\frac{E_0}{\bar{\varepsilon}}$ makes sense and coincides in order of magnitude with the ion free path until it stops or Residual Range, l_{tr} - ion transport mileage. For proton energies in the range: 1keV < E_0 < 30keV:

$$\sigma^* \sim 1. \tag{11}$$

Factors $(1-\Delta)^{5/2}$ and $(1-\Delta)$ in (9) take into account the dependence of elastic and inelastic cross sections on energy.

Function $e^{\left(-\frac{A}{\sqrt{1-\Delta}}\right)}$ in (9) is related to the phenomenon of charge exchange of reflected hydrogen ions [3,4], the constant A is an adjustable parameter.

Formula (9) was obtained under the assumption that there are no fluctuations in energy losses or other factors leading to a broadening of the energy spectrum. Since all broadening mechanisms: Doppler broadening, broadening associated with the non-monochromaticity of the probing beam and taking into account the instrumental function of the energy analyzer have a Gaussian profile, then taking into account all factors leading to the spread of energy losses can be determined on the basis of convolution:

$$S_{fin}(\Delta, \mu, \mu_0) = \int_0^{\infty} d\varepsilon \cdot S(\Delta - \varepsilon, \mu, \mu_0) \frac{1}{\sqrt{2\pi\sigma_\Sigma}} e^{\left(-\frac{\varepsilon^2}{2\sigma_\Sigma}\right)}, \tag{12}$$

where $\sigma_\Sigma = \overline{\varepsilon^2}s + \sigma_{Dop} + \sigma_{beam} + \sigma_{enan}$ - total broadening factor.

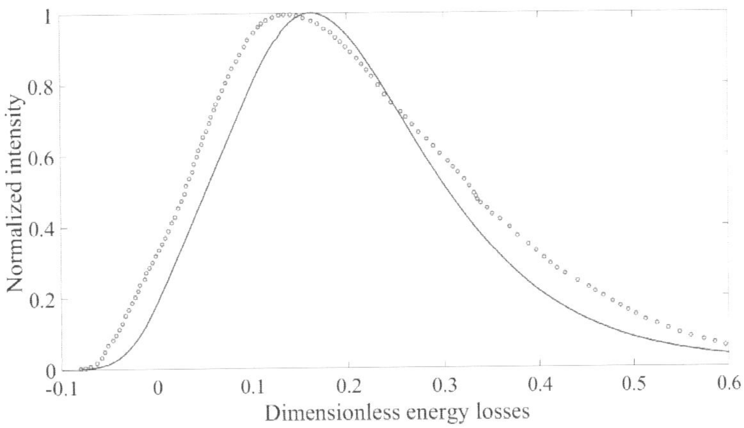

Figure 2. Energy spectra of protons with an initial energy of 25 keV, reflected from tungsten. Probing at an angle of 710, viewing angle - 1090, A = 1. Calculation – blue solid line based on formulas (7), (9) and (10), experiment – red circles [10].

Figure 2 shows the experimental [10] and calculated energy spectra of protons reflected from the tungsten array. As mentioned above, the elastic scattering parameters were determined on the basis of the KrC potential [1-3]. The protons stopping power is taken from the NIST tables. The change in stopping power from the value $\bar{\varepsilon}$ = 137.6 eV/nm, presented in the NIST data, to the value $\bar{\varepsilon}$ = 110.1 eV/nm, leads to the coincidence of the maxima of the calculated and experimental spectra.

The largest number of experimental data on the reflected ions energy spectra falls on the 80s and 90s of the last century [1–3, 10, 11]. Let us consider the results of experiments showing the dependence of the spectral width and the maximum on the target material.

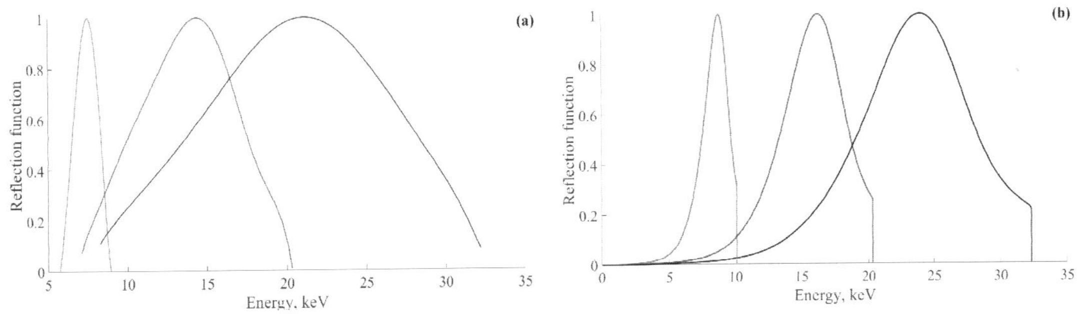

Figure 3. Energy spectra of protons reflected from copper: red solid line - 10.1 keV, blue solid line - 20.3 keV and black solid line - 33.2 keV. The incidence angle is 80°, the viewing angle is 110°. Part **a**: the experiment [12], part **b**: the calculation based on formula (9), $A = 1$.

Figure 3 illustrates the qualitative agreement between experimental data and calculations based on formulas (7) and (9). However an exact quantitative agreement is not achieved. The experimental maxima are shifted to the region of high energy losses. The main parameter influencing the position of the maximum is the stopping power $\bar{\varepsilon}$. Calculating the spectra in figure 3, we used $\bar{\varepsilon}$ data on taken from NIST. In order to achieve quantitative agreement with the data of [12], it is necessary to increase the value of $\bar{\varepsilon}$.

3. Conclusions and main results

Calculations based on the method presented in [6, 8] make it possible to describe Reflected Electron Energy Loss Spectrometry (REELS), Elastic Peak Electron Spectroscopy (EPES) and Reflected

Electron Spectra (RES) signals with satisfactory accuracy. The solution of the boundary problem for the elastic transport equation allows us to find the equation for the distribution of reflected atomic particles - Pass Length Distribution Function of reflected ions. Equation (2) obtained for PLDF has a universal and has the same form for both electrons and light ions, the difference concerns only the form of differential cross sections $x(\mu, \mu_0)$. It is shown that the maximum of the PLDF function (figure 2) is determined by the scattering geometry and weakly depends on the proton energy. In this work, the method presented in [6, 8] is used to solve (2) and describe the energy spectra of reflected light ions. It is shown that formulas (7) and (9) give a satisfactory description of the energy spectra of reflected protons. It is pointed out that it is inadmissible to use the SLA approximation to describe the energy spectra of reflected protons. Relying on the fact that the elastic scattering cross sections of light ions are much better studied than inelastic, then taking into account the sensitivity of the energy spectrum to the parameter σ^*, we obtain a technique that allows us to refine the values of $\bar{\varepsilon}$. Verification of the value $\bar{\varepsilon}$ is one of the most important problems in the physics of the light ions interaction with a surface.

Acknowledgments
The research was performed in National Research University «MPEI» thanks to the financial support of ministry of science and higher education of RF, state assignment № FSWF – 2020 – 0023.

References

[1] Mashkova E S and Molchanov V A 1980 *Rasseyniye Ionov Srednikh Energiy Poverkhnostyami Tverdykh Tel* (Moscow: Atomizdat)
[2] Kurnaev V A, Mashkova E S and Molchanov V A 1985 *Otrazheniye Legkikh Ionov ot Poverkhnosti Tverdogo Tela* (Moscow: Energoatomizdat)
[3] Mashkova E S and Molchanov V A 1985 *Medium Energy Ion Reflection from Solids* (Amsterdam: North-Holland)
[4] Ryazanov M I and Tilinin I S 1985 *Issledovaniye Poverkhnosti po Obtatnomu Rasseyaniyu Chastits* (Moscow: Energoatomizdat)
[5] Ziegler J F, Biersack J P and Littmark U 1985 *The Stopping and Range of Ions in Solids* (New York: Pergamon Press)
[6] Salvat-Pujol F and Werner W S M 2011 *Phys. Rev.* B **83(19)** 195416
[7] Afanas'ev V P and Naujoks D 1991 *Phys. Status Solidi* 133–40
[8] Afanas'ev V P, Efremenko D S and Kaplya P S 2016 *J. Electron. Spectrosc. Relat. Phenom.* **210** 16–29
[9] McCracken G M and Freeman N J 1969 *J. Phys. B (Atom. Molec. Phys.) ser. 2* **2** 661
[10] Eckstein W 1995 *Computer Simulation of Ion-Solid Interaction* (Berlin: Springer-Verlag)
[11] Bulgadaryan D G 2020 *Ph. D. thesis Rasseyaniye Protonov Kevnykh Energy kak Instrument Analiza Tonkikh Sloyev na Poverkhnosti Materialov* Moscow National Research Nuclear University MEPHI
[12] Kenji M, Hideo A and Tokuo S 1968 *Japan. J. Appl. Phys.* **7** 916

Electrochemical deposition of ceramic-like coatings on Ti-6Al-4V parts fabricated by electron beam melting

N Y Peretyagin[1,2], I V Suminov[1], A Smirnov[1], N I Krikheli[2], O V Kramar[2] and P Y Peretyagin[1,2]

[1] Moscow State University of Technology "STANKIN", 127055, Moscow, Russia
[2] Moscow State University of Medicine and Dentistry named after A.I. Evdokimov, 127473, Moscow, Russia

E-mail: n.peretyagin@stankin.ru

Abstract. The problems of micro-arc oxidation of titanium alloy parts Ti-6Al-4V produced by electron beam melting technology were considered. The phase composition, thickness and roughness of the obtained coating have been evaluated.

1. Introduction

Titanium and titanium alloys are widely used in aerospace, petrochemical, medical equipment field and other fields. Micro-arc oxidation (MAO), also called plasma electrolyte oxidation (PEO) or anodic-spark deposition (ASD), is a high-voltage anodic oxidation process in a plasma environment that is widely used for surface modification of valve metals. Valve metals are metals that can form adhesive, electrically insulating anodic oxide films after anodizing, such as aluminum, tantalum, zirconium, titanium, etc.

Additive manufacturing (AM) is an innovative manufacturing process. The essence of AM is to create components in almost any geometric shape without any subsequent machining, that allows to reduce production waste. The AM of titanium alloys includes two different methods: selective laser melting (SLM) and electron-beam melting (EBM). The main advantage of these methods as compared to traditional methods is that the product is created in one step, directly from computer aided design (CAD) model data [1, 2].

In particular, EBM is the most suitable AM method for chemically active metals such as titanium alloy because of their printing in high vacuum. In addition, EBM offers higher productivity (~80 cm^3/h), in contrast to SLM (20 - 30 cm^3/h), and more flexibility in the choice of powder particle size (20 - 45 μm with ELP, 45 - 105 μm with SLM) [3–7].

Purpose of work: to investigate the phase composition, thickness and roughness of the micro-arc oxide coating on the titanium alloy Ti-6Al-4V samples produced by electron-beam melting technology.

2. Material and methods

Samples of titanium alloy Ti-6Al-4V with dimensions of 28 mm x 19 mm x 5 mm were made on Arcam A2 machine by EBM of powder layers with thickness of 50 μm, average power of electron beam was 900 W, speed of the beam was 4350 mm/s. The entire process took place under vacuum conditions in a protective gas - helium.

Content from this work may be used under the terms of the Creative Commons Attribution 3.0 licence. Any further distribution of this work must maintain attribution to the author(s) and the title of the work, journal citation and DOI.
Published under licence by IOP Publishing Ltd

Micro-arc oxidation was carried out on an installation, allowing to set various electric modes of processing, in the bath filled with silicate-hypo-phosphite electrolyte from stainless steel. The process was carried out in soft anode-cathode mode at a total anode and cathode current density of 10 A/dm^2 and treatment duration of 10 minutes.

The phase composition was analyzed on an Empyrean diffractometer (PANalytical, Almelo, The Netherlands), in the Cu-Kα spectrum with a wavelength of 1.5405981Å, at 60 kV, a beam current of 30 mA, a diffraction angle 2θ ranging from 25° to 80° with a step size of 0.05.

The cross-sectional morphology of the samples was examined on a VEGA 3 LMH scanning electron microscope (Tescan, Brno, Czech Republic). The samples were ground and polished on a Tegramin-30 machine (Struers, Ballerup, Denmark).

Roughness evaluation was performed on the optical system MicroCAD premium+ (GFM, Berlin, Germany).

3. Results of investigations and discussion

Figure 1 shows diffractograms obtained from pristine and coated samples. The substrate contains two phases, alpha (α) and beta (β). After MAO the intensity of diffraction peaks of α- and β-Ti significantly decreases, and there is a tendency of complete disappearance of β-Ti. This is due to the fact that during the oxidation process the formed coating overlaps the initial matrix on the surface of the sample. The observation of the α-phase is due to the fact that the coating formed by MAO is thin and porous.

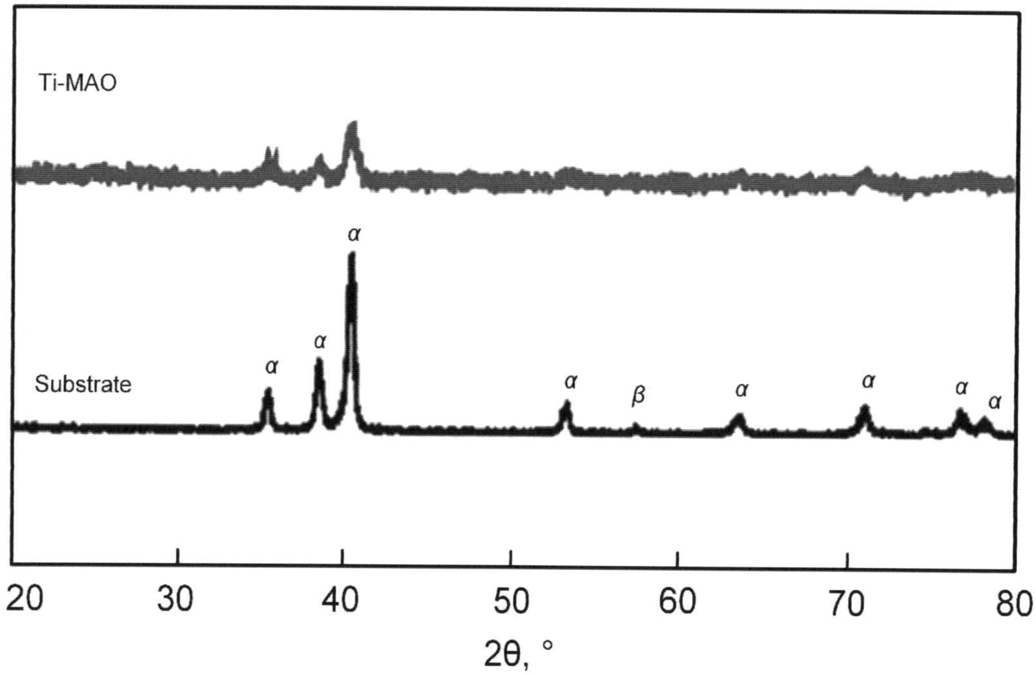

Figure 1. XRD patterns of pristine and coated samples (α: α-Ti, β: β-Ti).

The SEM cross-sectional view of the MAO coated sample is presented in figure 2. The average thickness of the coating was 3.4±0.32 μm. The substrate-coating interface is relatively smooth, which contributes to their adhesion.

Figure 2. The SEM cross-sectional view of the MAO coated sample.

Measurements of the surface roughness of the studied samples are presented in table 1. The results show that the surface roughness after the MAO process decreased (the sample without MAO coating: Ra=5.62 μm, Rz=32.12 μm; MAO coating roughness: Ra=1.82 μm, Rz=10.04 μm, respectively). The obtained low values of roughness are associated with uniform application, overlapping pores and irregularities obtained after the process of electro-beam melting.

Table 1. The results of determining the position of spectral lines.

			Ra, μm	Rz, μm
Ti-sample without MAO coating		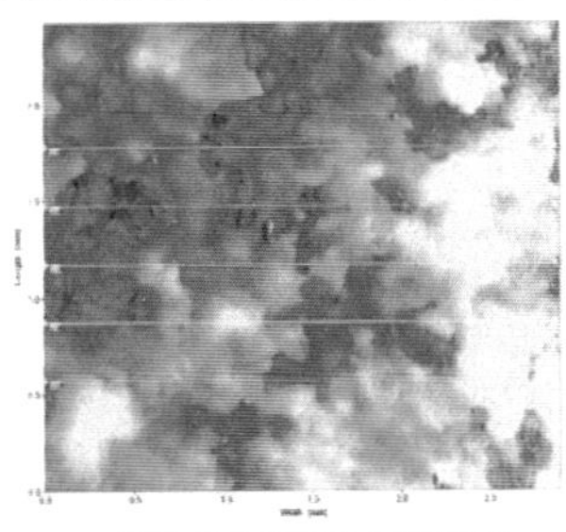	5.62	32.12
			Ra, μm	Rz, μm
Ti-sample with MAO coating		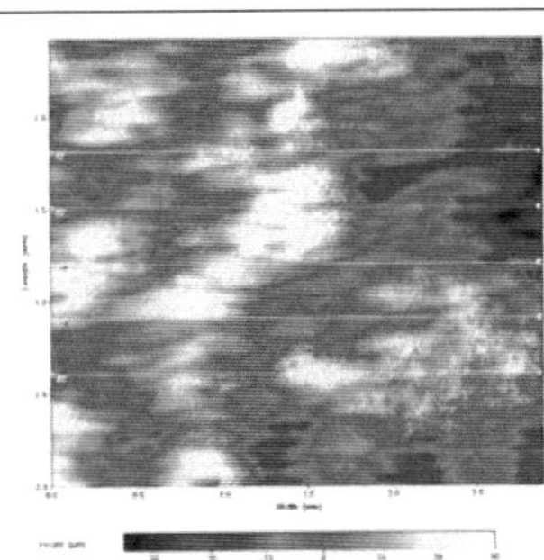	1.82	10.04

4. Conclusions

Samples of titanium alloy Ti-6Al-4V were fabricated by electron-beam melting, followed by MAO coating. Analysis of the phase composition of the coating showed that after MAO the intensity of diffraction peaks α- and β-Ti significantly decreases, and there is a tendency of complete disappearance of β-Ti. The thickness of the deposited MAO coating was 3.4±0.32 μm. Surface roughness of the deposited MAO coating had values Ra=1.82 μm and Rz=10.04 μm.

Acknowledgments

We would like to thank the Russian Science Foundation for supporting this work under the grant from the Russian Science Foundation (project No. 21-79-30058).

This work was carried on the equipment of the Collective Use Center of MSTU "STANKIN" (project No. 075-15-2021-695).

References

[1] Anikin K A, Borisov A M, Zheltukhin A V, Zhukov A A, Fedichkin I D, Apelfeld A V, Savushkina S V, Chernik V N 2018 *Journal of surface investigation: x-ray, synchrotron and neutron techniques* **12** 531–4

[2] Vereschaka A, Grigoriev S, Oganyan G, Tabakov V, Sitnikov N, Andreev N and Milovich F 2019 *Wear* **420-421** 17–37

[3] Vereschaka A A, Grigoriev S N, Vereschaka A S, Popov A Y and Batako A D 2014 *Procedia CIRP* 239–44

[4] Grigoriev S N, Melnik Y A, Metel A S and Volosova M A 2017 *Mechanics and industry* **18** 709

[5] Vereshchaka A A, Tabakov V P, Grigoriev S N, Sitnikov N, Andreev N and Milovich F 2018 *Wear* **416-417** 72–80

[6] Vereschaka A, Grigoriev S, Tabakov V, Migranov M, Sitnikov N, Milovich F and Andreev N 2020 *Tribology international* **150** 106388

[7] Kuzin V V, Grigoriev S N and Volosova M A 2014 *Journal of Friction and Wear* **35** 505–10

The effect of process current parameters on the properties of oxide layers under plasma electrolytic oxidation of AMg6 alloy

A V Polunin, A G Denisova, A O Cheretaeva, M R Shafeev, E D Borgardt, I A Rastegaev and A V Katsman

Research Division #4, Federal Government Budgetary Institution of Higher Education Togliatti State University, 445020, Togliatti, Russia

E-mail: anpol86@gmail.com

Abstract. The effect of current density and current ratio in the cathodic and anodic half-cycles during prolonged (180 minutes) plasma electrolytic oxidation (PEO) of AMg6 wrought alloy on the oxide layer wear and corrosion resistance were studied. It was established that the best wear resistance is achieved in the oxide layers obtained in the "soft sparking" mode (negative-to-positive pulse ratios of 1.15–1.30) at current densities of 9–15 A dm^{-2}, and the best set of wear resistance and corrosion resistance – in the oxide layers obtained in "symmetrical" mode (negative-to-positive pulse ratio of 1.00).

1. Introduction

Despite good plasticity-strength match of alloys of Al-Mg (magnalium) system [1], enhanced corrosion and wear resistance must be provided for their use in a number of applications. Furthermore, when using magnalia as casing material for portable devices and equipment (cell phones, laptops, various measuring instruments, etc.), there is also a need in aesthetic product qualities.

Presently, plasma electrolytic oxidation (PEO) is an effective and dynamically developing method of surface modification allowing to produce wear and corrosion resistant ceramic oxide layers on the surface of light non-ferrous metals, which increases the service life of products without changing their specific gravity significantly [2–4]. However, both the protective properties and the aesthetic qualities of the forming coatings strongly depend on the component composition of electrolyte and the parameters of process current for PEO [2, 4–6]. Unfortunately, theoretical prediction of the output properties of the coating based on the input data of the PEO process mode is not yet attainable [7]. Therefore, experimental study of the effects and optimization of PEO electrical modes is required to achieve the best consumer properties of the oxide layers formed [8–10].

In this paper, the influence of electrical current density and cathodic/anodic (C/A) current ratio in the C/A half–cycles was investigated during PEO of wrought AMg6 alloy specimens in an alkaline-phosphate-silicate electrolyte.

2. Experimental

Oxide layers were obtained on 20×60×6 mm^3 flat specimens of AMg6 magnalium (Al-Mg-Mn-Si system, 6.2 wt.% Mg). Before PEO, the specimens were ground on SiC 600-1500-2500-5000 grit abrasive paper, after which they were washed with isopropyl alcohol and distilled water. PEO was

Content from this work may be used under the terms of the Creative Commons Attribution 3.0 licence. Any further distribution of this work must maintain attribution to the author(s) and the title of the work, journal citation and DOI.
Published under licence by IOP Publishing Ltd

carried out in an electrolyte based on potassium hydroxide KOH (3 g/l), sodium pyrophosphate $Na_4P_2O_7 \times 10H_2O$ (8 g/l), and sodium metasilicate $Na_2SiO_3 \times 5H_2O$ (12 g/l) aqueous solution for 180 minutes. The values of current density j and *C/A ratio* varied in the range of 6–18 A/dm^2 and 0.70–1.30 in increments of 3 A/dm^2 and 0.15, respectively. At the same time, current frequency (500 Hz), duty cycle factor 65%, and interpulse time between cathodic and anodic impulses (1 to 1) remained constant.

The thickness and the structural features of the produced oxide layers were investigated on transverse sections using the scanning electron microscope (SEM Jeol JCM-6000 Neoscope). The specimen elemental composition was determined using energy dispersive X-ray fluorescence analysis (Shimadzu EDX-8000 spectrometer with Rh-anode tube). X-ray diffraction analysis of the specimens was performed in the Bragg-Brentano geometry in the angle range of 15-80° by 2θ, tube power (CuKα) of 1.6 kW (tube current – 40 mA, accelerating voltage – 40 kV), scanning speed of 0.25°/min, 0.02° step size on a Shimadzu Maxima XRD-7000 diffractometer. The oxide layer hardness HV was determined on sections on a Shimadzu DUH-211S microhardness meter at a load of 491 mN (10 measurements). Wear resistance of the obtained oxide layers in a friction pair with a steel counterbody (Ø 6.35 mm ball with hardness HRC 62...63) was evaluated using a universal Nanovea TRB 50N tribometer per the ASTM G133-99 Type A standard under 40 N load. The linear wear intensity of the friction pair (μm) was determined using the Macro Sensors CD 375 025 006 LVDT sensor. Corrosion resistance of the specimens was evaluated by the methods of potentiodynamic polarization and impedance spectroscopy in 3.5 wt% NaCl solution at room temperature (22±1°C) with the use of Elins P-45X potentiostat with an electrochemical impedance measurement module FRA-24M. Tests were performed in three-electrode pressure electrochemical cell CS936 (CorrTest®, Wuhan, China) with the locked area of the investigated specimen (1 cm^2). A 20×20 mm^2 platinum mesh was used as auxiliary electrode, and the potential of the working specimen was measured against the silver-chloride reference electrode (Ag/AgCl) filled with saturated KCl solution. Prior to the measurements, the specimens were held in the medium for 60 minutes for the open circuit potential (OCP) to stabilize. The specimen potentiodynamic polarization (PDP) was carried out in the range from -150 mV to +800 mV relative to OCP with potential sweep rate of 1 mV/s. The values of corrosion current density i_{corr} and corrosion potential E_{corr} were determined by Tafel curves. To perform electrochemical impedance spectroscopy (EIS), a sinusoidal disturbing signal with an amplitude of 10 mV was used in the frequency range from 10^{-2} Hz to 10^5 Hz. Experimental data were analyzed using the ZView software (Scribner Associates, USA).

3. Results

It was found that in PEO modes with a predominant anodic impulse, the increased porosity and cracking of the oxide layers are observed throughout the cross section, with individual pore channels and transverse cracks passing through the whole oxide layer down to the substrate (figure 1). PEO in the "soft sparking" mode [11] with C/A ratio = 1.15-1.30 brings about the formation of an almost defect-free oxide layer characterized by nearly total absence of transverse and main cracks, with significant porosity only in the outer zone of the oxide layer, which is subsequently removed (figure 1c).

The major phases in the all oxide layers are gamma-alumina (γ-Al_2O_3) and sillimanite (Al_2SiO_5), whose content ranges from 87.9 to 99.5 wt.% and from 0.5 to 12.1 wt.%, respectively. The fraction of amorphous phase in oxide layers ranges from 11.3 to 62.1 vol.%, and the greatest amorphous phase fraction (35–62 vol.%) is formed at high current densities (15–18 A/dm^2) in the "anodic" modes (with C/A ratio = 0.70–0.85). The decrease in current density to 6–12 A/dm^2 with simultaneous C/A ratio increase to 1.00–1.30 results in the amorphous phase fraction decrease to 11–30 vol.%.

The microstructure parameters (Scherrer's crystallite sizes, Stokes-Wilson residual microstresses) almost do not depend on the current density and C/A current ratio, and, apparently, are determined by the frequency and the duty cycle factor. The crystallite sizes of the γ-Al_2O_3 and Al_2SiO_5 phases, and their residual microstress levels ranged between 25–35 nm and 10–25 nm, and 0.34–0.43% and 0.48–1.42%, respectively, which is in good agreement with results in [12–14].

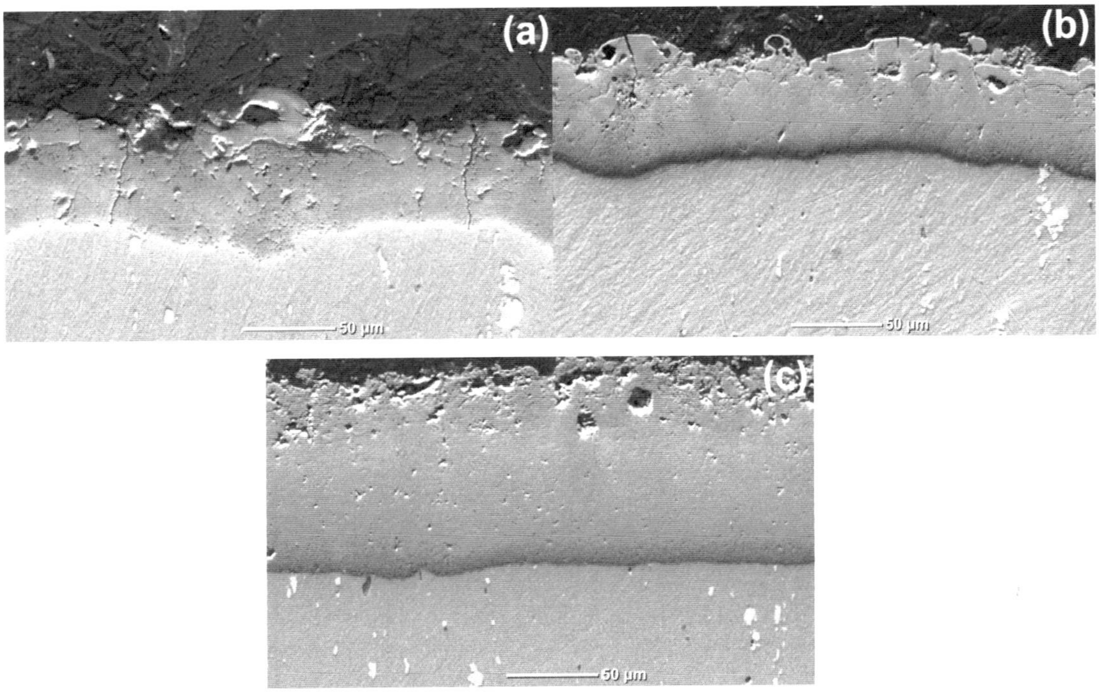

Figure 1. SEM images of transverse macrostructure of oxide layers obtained at current density 15 A dm-2 and different negative-to-positive pulse ratios 0.7 (a), 1.0 (b) и 1.3 (c).

It was found that current density has the greatest effect on the increase of the oxide layer thickness – with its increase from 6 to 18 A/dm^2 the thickness increases from 20–30 μm to 110–130 μm (Fig. 2). The C/A ratio has relatively little effect on the thickness (figures 1, 2), however, as noted above, largely determines the structure and crystallinity of the oxide layer, as well as its hardness, wear resistance and anti-corrosion properties, as will be demonstrated later.

For PEO in "anodic" modes (C/A ratio = 0.70–0.85), an increase in the current density from 6 to 9–12 A/dm^2 results in an increase in the hardness from 650–700 MPa to 950–1000 MPa, but further it stabilizes in this range. Increase in the C/A ratio to 1.00-1.30 results in the increase in the HV hardness to 1.3-1.45 GPa, and its highest values are achieved at low current densities (6–9 A/dm^2) in the "soft sparking" mode [11, 15]. In the same mode, the maximum wear resistance of the "oxide layer – steel indenter" friction pair was obtained, which is likely to be driven by the dominant abrasive nature of the wear of both bodies without coating chipping or formation of a large number of abrasive wear debris.

In all modes, PEO reduces corrosion current densities, i_{corr}, obtained by extrapolation of Tafel curves (ASTM G102), by 1-4 orders of magnitude: from $2{\cdot}10^{-7}$ A/cm^2 in pure alloy to, for example, $2.8{\cdot}10^{-11}$ A/cm^2 in the alloy with the oxide layer obtained at the current density of 12 A/dm^2 and C/A ratio = 0.85. Notably, still minimum values of i_{corr} normally cannot be achieved with PEO in the anodic modes. PEO with "symmetrical" C/A ratio = 1.00 for all current densities ensures one of the lowest and most stable i_{corr} values – (5…33)*10^{-10} A/cm^2. An increase in C/A ratio to 1.15–1.30 results in an increase in i_{corr} up to values (1.6…5.5)*10^{-7} A/cm^2 at PEO current densities of 15–18 A/dm^2, which are comparable to the i_{corr} value of the pure alloy. It should be noted that the minimum i_{corr} values are achieved at PEO current densities of 6–12 A/dm^2, while increasing up to 15–18 A/dm^2 leads to increased corrosion currents, which is probably due to the active formation of discharges and of pore channels up to the substrate during PEO.

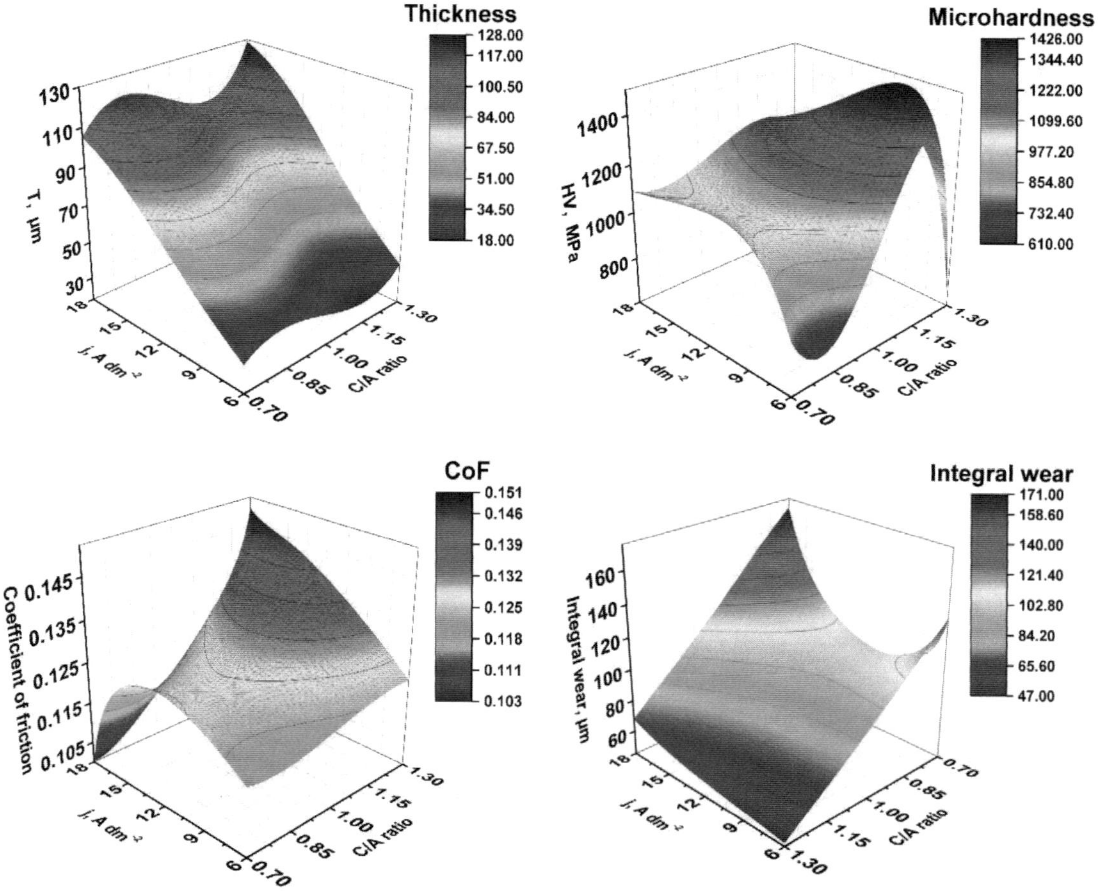

Figure 2. Thickness (T, μm), hardness (HV, MPa), friction coefficient and integral wear (μm) of oxide layers produced at different current densities (j, A dm-2), and negative-to-positive pulse ratios (C/A/ratio).

Impedance spectroscopy has shown similar trends in the corrosive behavior of magnalium AMg6. PEO allows to increase the impedance modulus $|Z|_{f=0,01\ Hz}$ up to 3 orders of magnitude - from $1.30 \cdot 10^5$ Ω*cm^2 in the pure alloy to $1,5 \cdot 10^8$ Ω*cm^2 in the alloy after PEO at C/A ratio = 0.85 and the current density of 12 A/dm^2. At the same time, the most stable and homogeneous $|Z|_{f=0,01\ Hz}$ indicators are achieved with PEO in "symmetrical" mode at the current densities of 6–15 A/dm^2. Further increase in C/A ratio to values 1.15–1.30 triggers a sharp drop of $|Z|_{f=0,01\ Hz}$ to the values (1.06…4.6)*10^5 Ω*cm^2, comparable to the value for a pure alloy. This experimental result is not fully understood and requires further research.

4. Conclusions

Thus, the optimal, most "balanced" characteristics of the oxide layers obtained by the PEO method are achieved when operating in the "symmetrical" PEO mode C/A ratio = 1.00 (i.e., at equal currents flowing through the specimen in the anodic and cathodic half-cycles). Deviation from this mode to the anodic side (C/A ratio = 0.70 ... 0.85) in some cases allows to achieve higher corrosion resistance, simultaneously leading to a decrease in the hardness and wear resistance of the oxide layer. C/A ratio increase to 1.15–1.30 results in the higher thickness, hardness and wear resistance of the oxide layers, but causes a parallel deterioration in their corrosion resistance.

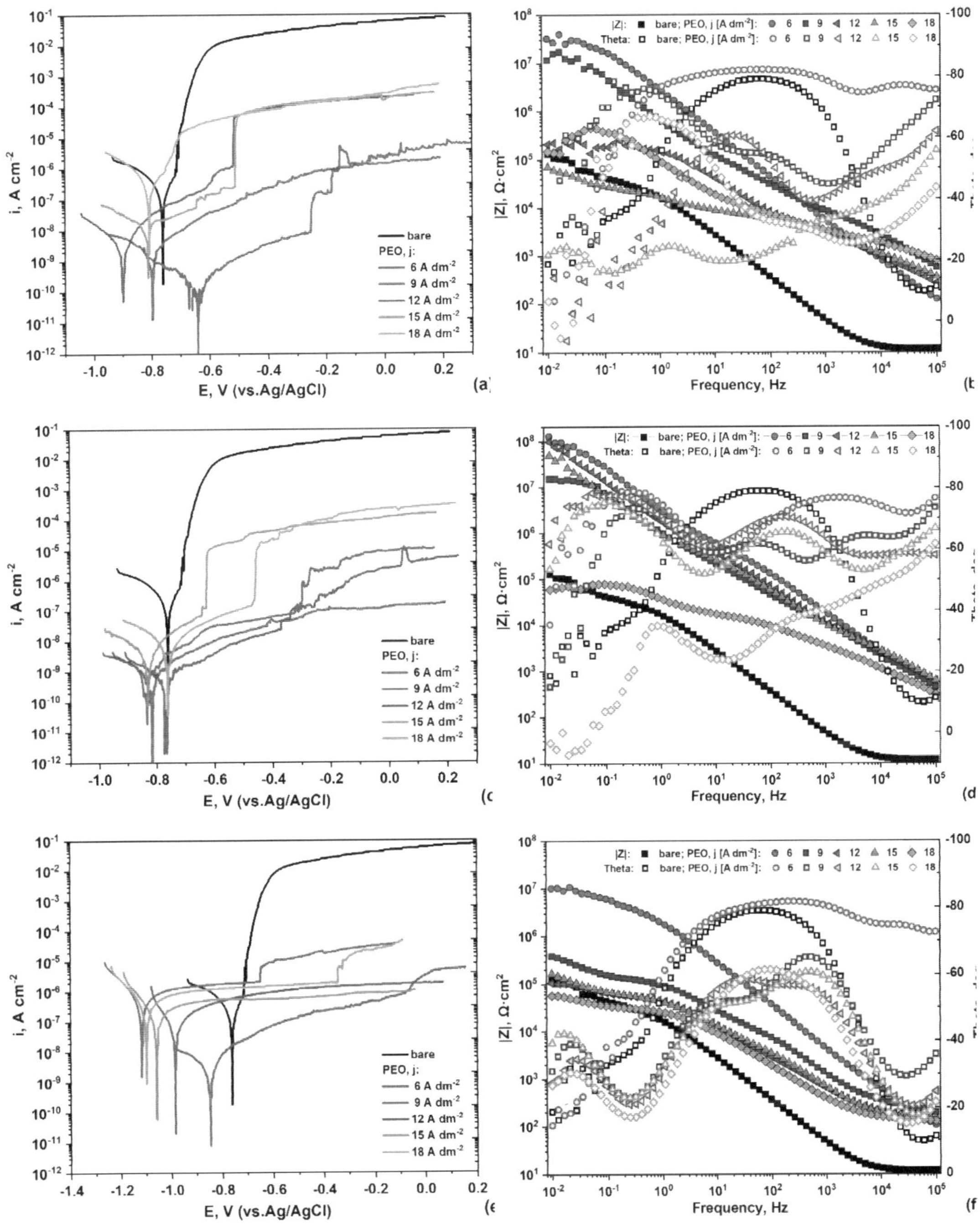

Figure 3. Tafel (left) and Bode (right) curves for initial alloy and for alloy after PEO with different current densities and negative-to-positive pulse ratios: 0.7 ((a),(b)); 1.0 ((c),(d)) and 1.30 ((e),(f)) in 3.5% NaCl.

It can be generally concluded that the PEO processing of the AMg6 PEO alloy makes it possible to effectively control its wear resistance and anticorrosive properties in a wide range of values.

Acknowledgements
The work was financed by a grant from the Russian Science Foundation (Project No. 20-79-10262). The authors are deeply grateful to Prof. Mikhail Krishtal (Togliatti State University) for his guidance and support.

References
[1] Lumley R N 2018 *Fundamentals of Aluminium Metallurgy* (Elsevier)
[2] Sikdar S, Menezes P V, Maccione R, Jacob T and Menezes P L 2021 *Nanomaterials* **11** 1375
[3] Simchen F, Sieber M, Kopp A and Lampke T 2020 *Coatings* **10** 628
[4] Yerokhin A L, Nie X, Leyland A, Matthews A and Dowey S J 1999 *Surf. Coatings Technol.* **122** 73–93
[5] Polunin A V, Borgardt E D, Shafeev M R and Krishtal M M 2019 *Journal of Physics: Conference Series* **1396** (Institute of Physics Publishing) 012032
[6] Lu X, Mohedano M, Blawert C, Matykina E, Arrabal R, Kainer K U and Zheludkevich M L 2016 *Surf. Coatings Technol.* **307** 1165–82
[7] Suminov I V 2011 *Plasma-electrolytic modification of the surface of metals and alloys* (Moscow: Technosphera)
[8] Erfanifar E, Aliofkhazraei M, Nabavi H F and Rouhaghdam A S 2017 *Surf. Coatings Technol.* **315** 567–76
[9] Fattah-Alhosseini A, Vakili-Azghandi M and Keshavarz M K 2016 *Acta Metall. Sin. (English Lett.)* **29** 274–81
[10] Kaseem M, Fatimah S, Nashrah N, Ko Y G, Gun Ko Y and Ko Y G 2021 *Prog. Mater. Sci.* **117** 100735
[11] Clyne T W and Troughton S C 2019 *Int. Mater. Rev.* **64** 127–62
[12] Curran J A and Clyne T W 2005 *Surf. Coatings Technol.* **199** 177–83
[13] Xie H Jun, Cheng Y Liang, Li S Xian, Cao J Hui and Cao L 2017 *Trans. Nonferrous Met. Soc. China (English Ed.)* **27** 336–51
[14] Polunin A V, Cheretaeva A O, Borgardt E D, Rastegaev I A, Krishtal M M, Katsman A V and Yasnikov I S 2021 *Surf. Coatings Technol.* **423** 127603
[15] Mohedano M, Matykina E, Arrabal R, Mingo B and Pardo A 2015 *Appl. Surf. Sci.* **346** 57–67

Study of nanocomposite hafnia and zirconia based layers produced by plasma spraying

S V Savushkina[1,2], A M Borisov[1,3], I V Suminov[3], E V Vysotina[2] and A A Ashmarin[2]

[1] Moscow Aviation Institute (National Research University), Volokolamskoe Sh. 4, Moscow, 121552, Russia
[2] Keldysh Research Center, Onezhskaya 8, Moscow, 125438, Russia
[3] Moscow State University of Technology "STANKIN", Vadkovsky per. 1, Moscow, 127994, Russia

E-mail: sveta_049@mail.ru

Abstract. Nanostructured and nanocomposite layers NiCoCrAlY+ ZrO_2-7%Y_2O_3, ZrO_2-7% Y_2O_3+HfO_2-9%Y_2O_3, HfO_2-9%Y_2O_3 with thickness of ~ 20 μm were formed by low pressure plasma spraying. The structure and composition of the layers have been studied using a scanning electron microscopy, X-ray microanalysis, and XRD analysis. Thermal stability of the coatings has been analyzed using synchronous thermal analysis at temperatures up to 1600 °C. The results of structure and composition analysis of ZrO_2-7%Y_2O_3+HfO_2-9%Y_2O_3 layer suggest the formation of nanocomposite co-doped regions of the ZrO_2-HfO_2-Y_2O_3 solid solution. The layer has greater thermal stability at temperatures up to 1600 ° C and a smaller subgrain size (~ 33 nm) than for the ZrO_2-7% Y_2O_3 and HfO_2-9%Y_2O_3 layers.

1. Introduction

Protective nanostructured coatings based on such high-temperature materials as zirconium and hafnium oxides are relevant both for nuclear power, rocket and space technology [1–5]. One of the problems of nanostructured materials use is their low thermal stability. With an increase in the working temperature, grain growth occurs and the structure becomes coarser, which can eliminate the effect of improving the functional properties. In addition, a disadvantage of materials based on stabilized zirconium and hafnium oxides is the partial or complete evaporation of rare earth oxides, which stabilize the cubic lattice at high temperatures [6, 7]. Composite and complex alloyed structures of coatings improve their thermal and chemical stability, mechanical strength, and reduce thermal conductivity [8–10]. New promising classes of materials are high-entropy and medium-entropy oxides [7, 11]. Usually these are single-phase oxide systems containing 3-5 or more cations. The creation of a high-entropy structure can make it possible to expand the operation temperature range of nanostructured materials. Entropy stabilization of multicomponent oxide solid solutions improves the performance of materials, including at high temperatures. The use of triple medium-entropy systems in coatings will expand the stability temperature-concentration range of fluorite-like solid solutions based on zirconium and hafnium dioxides. Medium-entropy oxides, in comparison with low-entropy binary oxide systems, are characterized by weaker diffusion, stronger distortion of the crystal lattice, and "cocktail" effects. These features lead to a slowdown in grain growth, preservation of the grain

Content from this work may be used under the terms of the Creative Commons Attribution 3.0 licence. Any further distribution of this work must maintain attribution to the author(s) and the title of the work, journal citation and DOI.
Published under licence by IOP Publishing Ltd

nanoscale in the coating, phase stabilization, an increase in thermal stability, and a decrease in thermal conductivity [7, 11].

2. Experimental setup and characterization techniques

Nanostructured and nanocomposite layers were obtained on Cu (99.9%) samples with a diameter of 30 mm and a thickness of 2 mm by low pressure plasma spraying (chamber pressure of ~ 100 Pa). The consumption of the plasma-forming gas (nitrogen) was 0.85 g/s, the powder consumption was 0.2 g/s. The arc current was ~ 120 A, the voltage was ~ 60 V. DC plasma gun equipped a supersonic nozzle with two turning sections creating Prandtl-Maier expansion fan was used for coatings spraying [12]. When the plasma flow with the sprayed material rotates in turning sections, a rarefaction waves fan forms, and the temperature drop rate increases (~ 10^8 K/s). This effect leads to intense condensation of nanoparticles from the vapor phase [10, 13]. We used NiCoCrAlY, ZrO_2-7%Y_2O_3 (YSZ), HfO_2-9% Y_2O_3 (YSH) powders with a particle size in range of 10-40 μm, as well as a mixture of NiCoCrAlY + YSZ and YSZ + YSH powders obtained in a ball mill for spraying. Nanostructured (YSZ, YSH) and nanocomposite layers (NiCoCrAlY + YSZ, YSZ + YSH) ~ 20 μm thick have been formed. The spraying process was carried out by reciprocating movement of the plasma gun along the rotating holder with the samples.

The structure of the layers was investigated using a Quanta 600 scanning electron microscope (SEM) with a TRIDENT XM4 microanalysis system and an X-ray diffractometer (Empyrean PANalytical, CuK radiation). The thermal stability of the initial powders and sprayed coatings was investigated using a NETZSCH STA449F 1 thermal analyzer (Jupiter) by differential scanning calorimetry (DSC) and thermogravimetric analysis (TG) methods in corundum crucibles in the temperature range 40-1600 °C at a rate of 5°/min. The protective and purge gases were argon with the flow of 50 ml/min and air with a flow of 50 ml/min.

3. Results and discussion

The highest content of nanofraction and the smallest particle size in the structure were obtained for the cermet coating NiCoCrAlY + YSZ. The YSZ coating contains both nanostructured regions and large particles. The structure of YSZ and YSZ + YSH coatings is similar. It mainly consists of globular particles ranging in size from 0.06 to 0.3 μm (figure 1, a). X-ray microanalysis of the YSZ + YSH coating showed a predominance of mixed composition areas. Figure 1, b shows the values of the configurational entropy calculated using the microanalysis data at 15 different points of the surface. The obtained values of the configurational entropy are predominantly in the range 1.5 R> $S_{config} \geq 1$ R, which suggests the formation of nanocomposite co-doped regions which have the properties of medium-entropy oxides. This regions contain ZrO_2-HfO_2-Y_2O_3 solid solution with a fluorite structure, characterized by mixing of Zr^{4+} and Hf^{4+} ions.

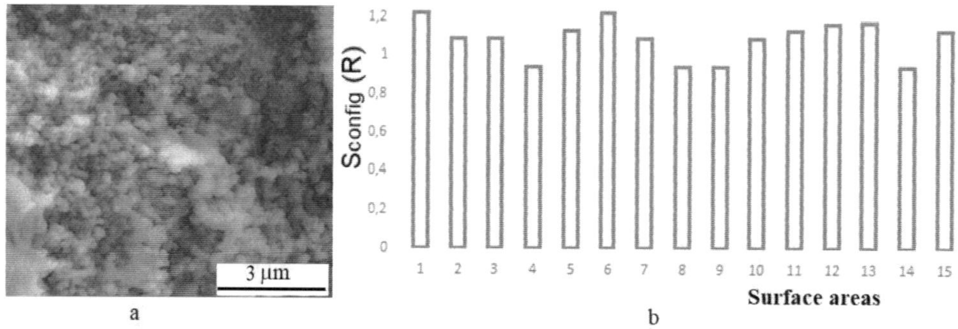

Figure 1. The surface structure of the YSZ+YSH coating (a) and the configurational entropy calculated from the data of X-ray microanalysis in different areas of the coating (b). R is the universal gas constant.

According to the data of synchronous thermal analysis, the thermal stability of the initial powders is generally higher than for the corresponding sprayed coatings. This difference is associated with the presence of oxygen vacancies, chemisorbed hydroxyl groups, and an amorphous and nanocrystalline component in the coatings. The total weight increase in the investigated temperature range for the cermet coating NiCoCrAlY + 7YSZ is ~ 16%. In NiCoCrAlY, there are intensive processes associated with the oxidation of Co, Ni, Al, the formation of spinels $NiAl_2O_4$, $CoAl_2O_4$, $NiCr_2O_4$, chemical reactions between oxides Al_2O_3, Cr_2O_3, NiO with the formation of spinel $Ni(Cr, Al)_2O_4$, etc.

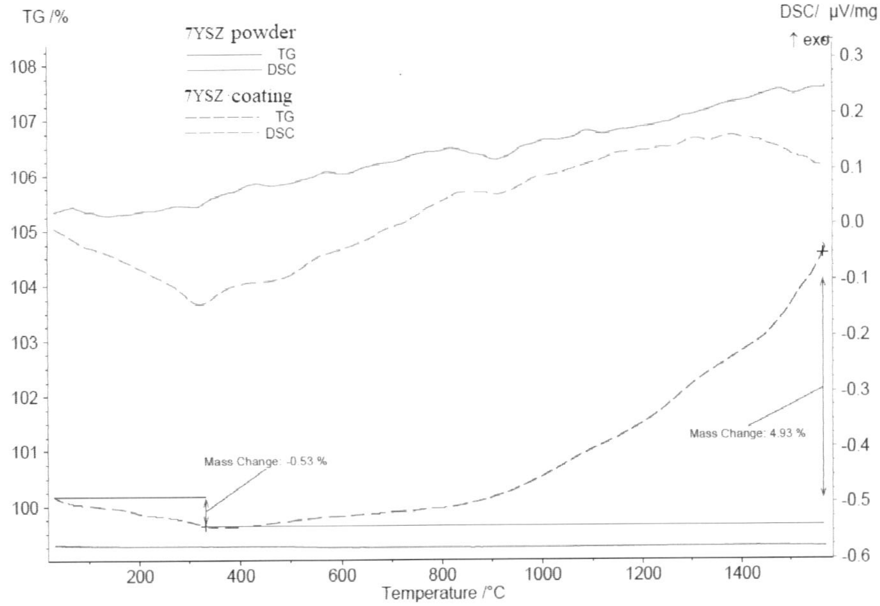

Figure 2. Graphs of differential scanning calorimetry (DSC) and thermogravimetric analysis (TG) on temperature for YSZ coating and initial powder.

DSC and TG curves of ceramic coatings have a more flat form. Wide endothermic peaks below 300°C may be due to the release of gases from the coating materials (water vapor, CO_2). Weak exothermic peaks at temperatures of 800 - 1000°C, accompanied by an increase in mass, can be associated with the oxidation of residual carbon on the coating surface. The presence of DSC peaks at temperatures above 1000 ° C can indicate the processes of the amorphous component crystallization, grain growth, and reduction of non-stoichiometric oxide in the coating. The YSH + YSZ layer has the highest thermal stability. The change in mass in the range 1200-1600 °C is less than 0.7% (figure 2), ~ 1.2% for YSH and the largest for YSZ ~ 5% (figure 3). Thermal analysis of the initial powders YSH, YSZ, YSH + YSZ showed the absence of an exothermic process and weight loss at a temperature of about 1400 °C. The peaks of weight loss in the temperature range from 1300 to 1600°C can be associated with the release of chemisorbed hydroxyl groups. As a result of exposure to high temperatures during plasma spraying in a vacuum, oxygen can be partially removed in the oxide coating. Oxygen vacancies in the surface layer of oxide coating can act as active catalytic sites for adsorbate binding. When the surface of the oxide coating was exposed to the atmosphere, this could lead to the formation of a direct product of water dissociation in oxygen vacancies — paired hydroxyl groups, as noted in [14]. In [15], experiments on heating showed that for the complete removal of hydroxyl groups from the plasma synthesized hafnium oxide particles, a temperature of more than 1200°C is required.

The YSH coating is characterized by a cubic phase, while YSH + YSZ and YSZ are crystallized in tetragonal phase. The subgrain size has been provided by coherent scattering region analysis in X-ray diffraction. The smallest subgrain size was found for the YSZ + NiCoCrAlY (210 Å) and YSZ + YSH (330 Å) nanocomposite layers, and higher for the YSZ and YSH layers.

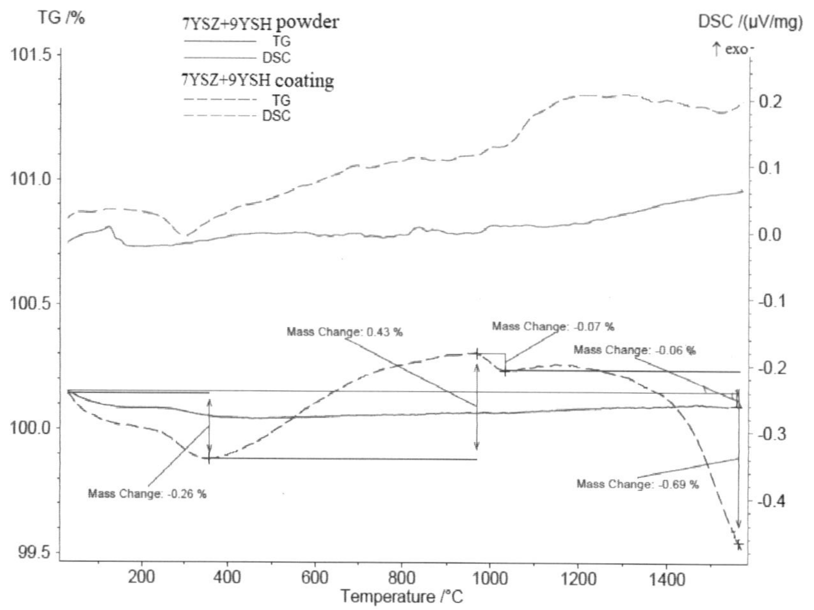

Figure 3. Graphs of differential scanning calorimetry (DSC) and thermogravimetric analysis (TG) on temperature for YSZ +YSH coating and initial powder.

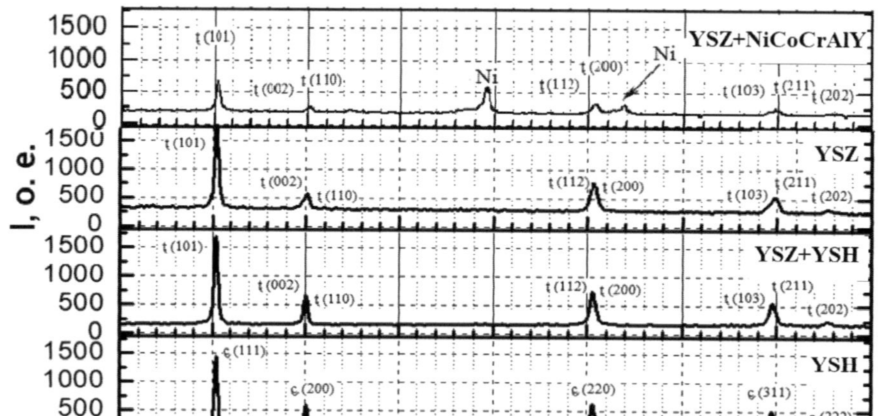

Figure 4. XRD spectra of the sprayed layers.

The structure of the ceramic coating of the mixed composition YSZ + YSH has showed both higher thermal stability at temperatures up to 1600 ° C and a smaller subgrain size compared to YSZ and YSH. Such features can be associated with the presence of the ZrO_2-Y_2O_3-HfO_2 solid solution regions, which promote to prevent grain growth and phase transitions, as occurs in medium-entropy and high-entropy systems.

4. Conclusions

Nanocomposite (YSZ + YSH, YSZ + NiCoCrAlY) and nanostructured (YSZ, YSH) layers were obtained by plasma spraying in a vacuum chamber using a nozzle equipped with a turning section. The highest content of nanofraction in the structure was obtained for the cermet coating NiCoCrAlY + YSZ. Calculated estimates and the results of studying the YSZ + YSH coating structure suggest the formation of nanocomposite complex doped regions of the ZrO_2-HfO_2-Y_2O_3 solid solution with a defective fluorite structure characterized by mixing of Zr^{4+} and Hf^{4+} ions. This system has the properties of medium-entropy oxides, which is confirmed by higher thermal stability at temperatures up to 1600 ° C, and a smaller subgrain size (~ 33 nm) than for the YSZ and YSH layers.

Acknowledgments
This research was carried out with the financial support of the Russian Science Foundation within the framework of scientific project No. 21-79-30058.

The study was carried out using the equipment of the Center for Shared Use "State Engineering Center" MGUT "STANKIN" supported by the Ministry of Higher Education of the Russian Federation (project 075-15-2021-695).

References

[1] Noor-A-Alam M, Choudhuri A R and Ramana C V 2011 *Surf. Coat. Technol.* **206** 1628–33
[2] Thibblin A, Jonsson S and Olofsson U 2018 *Surf. Coat. Technol.* **350** 1–11
[3] Yang D, Gao Y, Liu H and Sun C 2017 *Surf. Coat. Technol.* **315** 9–16
[4] Pourbafrani M, Shoja Razavi R, Bakhshi S R, Loghman-Estarki M R and Jamali H 2015 *Surf. Eng.* **31 (1)** 64–73
[5] Gong W B, Sha C K, Sun D Q and Wang W Q 2006 *Surf. Coat. Technol.* **201** 3109–15
[6] Sevastyanov V G, Simonenko E P, Simonenko N P, Stolyarova V L, Lopatin S I and Kuznetsov N T 2015 *Mat. Chem. Phys.* **153** 78-87
[7] Hachél M J R, Cheng C and Zou Y 2020 *J. Mat. Res.* **35** 8 1051–75
[8] Vereschaka A S, Grigoriev S N, Tabakov V P, Sotova E S, Vereschaka A A and Kulikov M Y 2014 *Key Eng. Mater.* **581** 68–73
[9] Vereschaka A, Tabakov V, Grigoriev S, Sitnikov N, Oganyan G, Andreev N and Milovich F 2019 *Wear* **420-421** 17–37
[10] Savushkina S, Polyansky M, Vysotina E, Ashmarin A and Tkachenko N 2019 *Surf. Coat. Technol.* **361** 212–21
[11] Sarkar A, Wang Q, Schiele A, Chellali M R, Bhattacharya S S et al. 2019 *Adv. Mat.* **31** 1806236
[12] Dumitrescu O, Gherman B and Dragan V 2018 *IOP Conf. Ser.: Mater. Sci. Eng.* **400** 042017
[13] Polyanskiy M N and Savushkina S V 2014 *J. Surf. Invest.: X-Ray, Synchrotron Neutron Tech.* **8** 144–48
[14] Choudhury B, Chetri P and Choudhury A 2015 *J. Exp. Nanosci.* **10** 103–14
[15] Forker M and Presa P 2008 *Phys. Rev.* **B 77** 054108

Regularities and features of acoustic emission under plasma electrolytic oxidation of wrought Al-Mg alloy

I A Rastegaev and A V Polunin

Federal Government Budgetary Institution of Higher Education Togliatti State University, 445020, Togliatti, Russia

E-mail: RastIgAev@yandex.ru

Abstract. The paper highlights the characteristic features of acoustic emission (AE) in the process of plasma electrolytic oxidation (PEO) of aluminum alloy, and also reveals the correlation of 4 main stages of PEO process with AE signals. The fundamental possibility to establish and detail the features of oxidation stages and to compare different PEO modes by AE signals was demonstrated. The results obtained substantiate the high potential of AE method as an instrument of in situ research, production monitoring and control, and evaluation of quality of PEO process on aluminum alloys.

1. Introduction

Plasma electrolytic oxidation (PEO) is one of the main and most promising methods of electrochemical thermal treatment of non-ferrous alloys (Al, Mg, Ti, Zr, Nd, etc.) allowing to produce "coatings" (oxide layers) with high mechanical, anticorrosive, thermal insulation and electrical insulation properties on the working surfaces of parts [1, 2]. However, the modern approach to the development of PEO technology and its adaptation to specific technical tasks normally consists of the empirical selection of process current parameters and electrolyte component compositions in order to achieve the best properties of the forming oxide layers that meet the product requirements. In this regard, most of the studies were carried out in different (incomparable and non-reproducible) conditions – different electrolytes, electrical processing modes, different treated alloys. Therefore, deriving common regularities that allow predicting the effect of a given variable factor on the oxide layer properties, or determining the optimal parameters to achieve the best oxide layer properties cannot be accomplished.

Many authors noted [2–6] that in order to obtain high-quality PEO oxide layers on non-ferrous alloys it is necessary to monitor the oxidation process in situ at the structural level. Concurrently, the problem is most vital for PEO of casing-type products with a complex shape (cylinder blocks, compressor housings, etc.), for which several working surfaces are treated in one process stage. However, methods and technical solutions for in situ monitoring and adjustment of the PEO process are still far from real practical applicability and require research and improvements at the level of devices and systems.

One of the promising methods for PEO mode monitoring may be the acoustic emission (AE) method [7]. Yet a very limited number of papers are devoted directly to AE diagnostics of PEO [3, 6] and were successful in proving experimentally that there is a primary correlation of AE nature with PEO electrical mode. Presently, there is almost no understanding of the nature and features of AE at

Content from this work may be used under the terms of the Creative Commons Attribution 3.0 licence. Any further distribution of this work must maintain attribution to the author(s) and the title of the work, journal citation and DOI.
Published under licence by IOP Publishing Ltd

the main stages of oxide layer formation in PEO on various alloys. Therefore, the purpose of this work was to study the patterns and features of AE during PEO of wrought Al-Mg alloy to assess the possibility of creating in situ monitoring and diagnostic control tools. This will make it possible in the future to create highly sensitive precision control systems for PEO modes for the formation of oxide layers on products of complex shape.

2. Experimental materials and methods

It is known that the recorded AE signal has minimal distortion only near its source [7], therefore, AE sensor should be placed directly on the specimen, and as close as possible to the region of actual process flow. However, the fulfillment of this condition during PEO is difficult, because of the high voltage and electrical current flowing through the electrolyte between the specimen and the electrode. Thus, without a special auxiliary circuit, such measurements will inevitably result in damage to expensive AE equipment. Therefore, before performing this work, a special measurement circuit for registering AE from a specimen with three protective barriers was developed, its essential follows below. Protective barriers should ensure the safety of analog-to-digital converter board (ADC) of the information processing unit (IPU) as the most expensive element of the AE equipment. The dielectric layer between AE sensor (AES) and the specimen was used as the first protective barrier. It was also the AES holder and the acoustic couplant. The second barrier was an interface device (ID) installed immediately at the AES output. The task of using the ID also includes minimizing the losses when transmitting a signal to the external preamplifier (PA) located in a safe zone. An external PA with adjustable gain is the third barrier. Thus, the dielectric acoustic couplant was the protection cornerstone. Should its breakdown occur, inexpensive ID will burn out. However, since it is located almost directly at AES within the electrode bath area, the connecting cable is likely to fail. In this case, and also if the breakdown passes through ID, the external PA burns out. The power supply of intermediate devices (ID and PA) is low-voltage and independent (from external rechargeable batteries). No-filler epoxy glue was used as a dielectric acoustic couplant; AES is band piezoelectric sensor П111-(0.02-0.3) (Russia); PA is 2-cascade broadband filter amplifier MSAE-FA010 (Russia); ADC is 4-channel 12-bit Advantech PCI-1714 board (Taiwan); ID is 2-cascade broadband preamplifier with zero gain (own make). The results of approbation demonstrated the operability of this measurement circuit with the possibility of amplifying acoustic signals without increasing the level of electrical crosstalk that unavoidably accompany measurements during PEO and that prevailed over the level of AE signals in other measurement circuits reviewed (AE registration through a bath [3]; through the air with a microphone [6]; through a waveguide welded to the specimen; immersionally, by immersion of the insulated AES in a bath). The operability of the proposed measurement circuit was confirmed using a calibrated acoustic pulse generator MSAE-UCA01 (Russia) according to technique [8].

In the process of PEO, AE was recorded continuously using a standard ADC recording program in the range of 0.02-1.0 MHz with a sampling frequency of 2 MHz. However, given the gradual changing of PEO process, AE recording was carried out in frames lasting 40 ms (20 full cycles of exposure) every 2 minutes, which is quite enough to trace any fast (the result of each exposure cycle) and slow (the result of several exposure cycles) changes in the process. Frames of voltage and current changes in the same PEO cycles were recorded synchronously with AE. Comparison and processing of research data were performed using the Octave mathematical package.

Specimens for PEO were made in the form of 200×20×6 mm strips of wrought Al-Mg alloy 1560 (chemical composition, wt%: 6.2 Mg; 0.65 Mn; 0.5 Ti, 0.4 Si; 0.3 Fe; 0.18 Zn; 0.087 Cu; Al – bal.), which is one of the main structural alloys in carriage-, ship-, machine-building industries, etc. PEO of specimens was carried out at alternating current (AC) at the actual surface area 34±1 cm^2, in an electrolyte based on distilled water aqueous solution with addition of KOH (3 g/l), sodium pyrophosphate $Na_4P_2O_7 \times 10H_2O$ (8 g/l), and sodium metasilicate $Na_2SiO_3 \times 5H_2O$ (12 g/l). All reagents were of chemical purity. Constant temperature of the electrolyte was maintained at 285±3 K with the use of the cooling system during PEO.

The research included investigations of AE under several modes of PEO differing in: oxidation process duration (60, 90, 120, 150 and 180 minutes); current density (J = 6, 9, 12, 15 and 18 A/dm^2), and negative-to-positive pulse ratio (C/A ratio, C/A = 0.7, 0.85, 1.0, 1.15 and 1.3). A total of 125 variants of PEO modes have been studied. For all PEO modes, current frequency (500 Hz), process current duty cycle factor (65%), cathodic and anodic forming pulse duration ratio (55%), and interpulse time between cathodic and anodic impulses (50%) were fixed.

3. Results and discussion

In the framework of this work, the results of experiments with the maximum PEO duration (180 min) are provided as most complete examples including all stages of the PEO modes studied. Notably, the following derivations, regularities and features of AE signals in PEO are fully valid for the shorter PEO durations.

The main stages of oxide layer formation during PEO of aluminum and magnesium alloys are known to be distinguished visually, and using photo and video recording [5] (figure 1), or photoelectric effect registration [5, 8], or by changing the peak voltage (anodic forming A. Gunterschulze curves), or by the voltage curve changing shape during the exposure cycle [2, 4, 5]. The main PEO stages are the following (figure 1):

I - anodizing and the beginning of anodic "sparking" (AS), which is visually accompanied by a general glow of the specimen;

II - PEO with predominantly anodic microarc discharges (AMAD), in which uniform and merging micro-flashes of white or blue-white color are visually observed on the overall dark background of the specimen;

III - transition (mixed) mode of PEO with the occurrence of microarc discharges in the anodic and cathodic half-cycles (ACMAD), in which a visual pattern similar to stage II is observed, but micro-flashes are more rare and powerful (bright), and with the beginning shift of their color to a white-yellow tint;

IV - PEO mainly with cathodic polarity including the formation of local arc discharges (AD) visually representing a "soft glow" from inside the specimen, mainly of yellow or yellowish-red tint.

The established optical and electrical signs of PEO stages were also observed in our experiments and were used to determine timing marks of the beginning and the end of the defined oxidation stages I-IV. Based on the timing marks, it was revealed that at each PEO stage, a characteristic AE type corresponds to cycles of electrical voltage, which is illustrated by the example of a mode with a current density of 6 A/dm^2 and a current ratio C/A=1.15 (figure 1) as including all four characteristic stages of the PEO process.

Figure 1 shows that stage I is characterized by continuous AE (a large number of burst signals of comparable amplitude) in the anodic part of the exposure cycle and almost no AE in its cathodic part, which becomes clear from the chronological recording frames of voltage and AE amplitude changes in one full exposure cycle. At stage II, amid a continuous AE, higher amplitude distinguishable AE burst signals appear in the anodic part, and AE signals also appear in the cathodic part of the cycle. At stage III, AE signals are also present in the anodic and cathodic parts of the cycle, but at the anodic half-cycle it is mainly formed by burst acoustic signals. At stage IV, AE signals virtually disappear in the anodic exposure half-cycle (they become low-amplitude) and are observed after powerful AE signals appear in the cathodic half-cycle. However, over time, the repetition frequency of AE signals noticeably decreases, i.e. AE does not occur in every cycle of exposure. From the same recording frames, the shape of voltage curve is seen to be changing at each PEO stage, which is consistent with the data [2, 4]. Furthermore, by peak voltage values, the result of determining the stages also corresponds to the inflection points of the A. Gunterschulze anodic forming curves, built with account for the high voltage region [2, 4].

Thus, by comparing AE to exposure cycles, the stages of PEO can be recognized and the times of their beginning and end can be determined. Further, using this AE information, an general comparison (evaluation) of various PEO modes can be performed by the stages achieved in each mode, and also by the duration of processing at each stage. The latter is illustrated by the example shown in figure 2, in which AE records (similar to the one provided in figure 1) obtained when accompanying different PEO modes with the same process duration (180 min) are summarized.

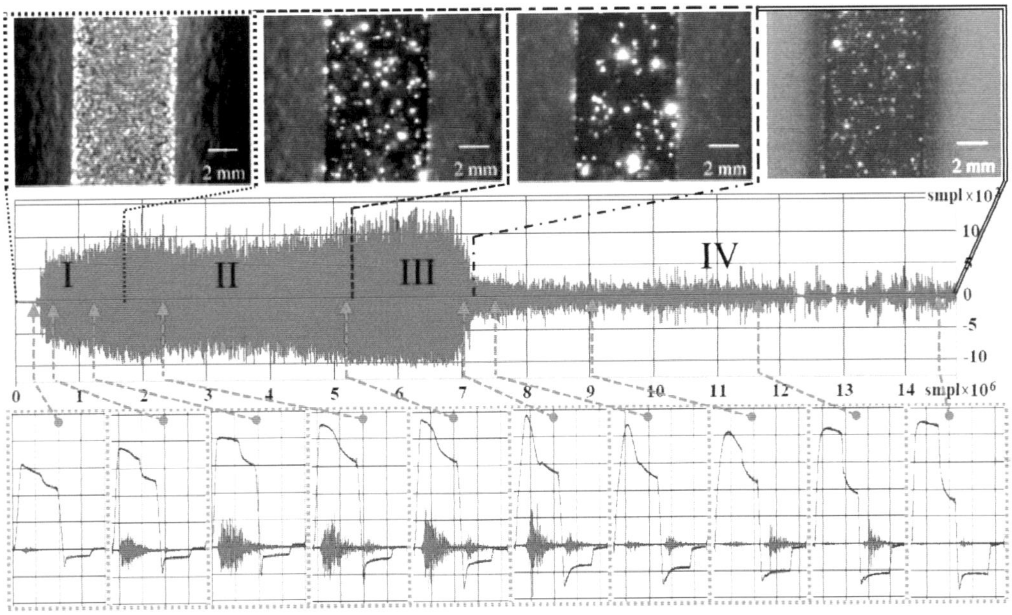

Figure 1. General pattern of AE amplitude change staging in cycles, where: I - IV are PEO stages; photographs at the top are visual signs of PEO stages per [5]; inserts at the bottom are chronological record frames of AE (red line) and voltage (blue line) in the exposure cycle; amplitude and time are presented in ADC samples (smpl).

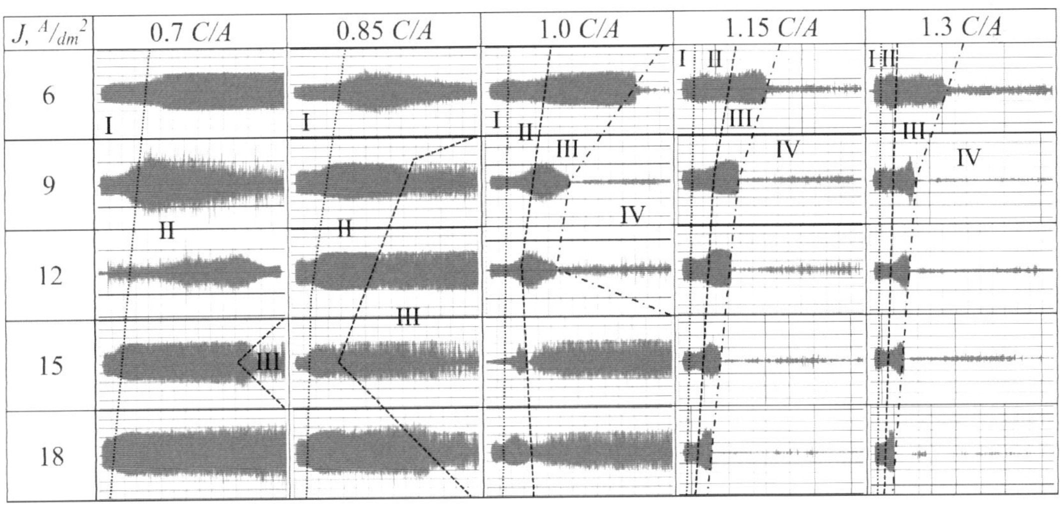

Figure 2. The general pattern of AE amplitude change depending on the PEO mode with oxidation duration of 180 minutes (designation of AE stages as in figure 1; designations of the control parameters are transcribed in the text).

In figure 2, the boundaries of PEO stages are plotted as per AE signs in accordance with the foregoing data and which are also confirmed in two ways: firstly, consistency with the visual signs of PEO stages; secondly, consistency with the change in PEO voltage. In addition, the indicated AE behavior is in line with modern concepts of PEO sequence [1, 2] according to which the process begins (stage I) with primary passivation and anodizing of the surface across the entire area of the specimen immersed in the electrolyte as a result of electrochemical reactions. In this case, there are many surface AE sources of low power, signals from which overlap with each other. Therefore, it is logical that their resulting signal is of a noise-like (continuous) type. Then, at stage II, an electrical breakdown of the forming thin passive (barrier) layer begins, which is accompanied by high power acoustic pulses in the anodic half-cycle. As the oxide layer grows and the voltage increases, the occurrence of oxide layer micro-arc breakdown begins in the both the anodic and the cathodic half- cycles (stage III). At the same time, many authors note that cathodic breakdown is lower in power and occurs at the metal substrate (in the coating depth), which is unfavorable as regards its capturing by optical methods [5, 8]. However, this is beneficial for the AE method, since the signal is actually generated on an acoustic waveguide (specimen), on which the AE sensor is directly installed. Atstage IV, the process almost completely goes "inside the coating" that has a high resistance to propagation of AE signals (due to scattering and attenuation of waves inside the ceramic oxide layer and reflection at the substrate interface). Therefore, it is accompanied by a decrease in the amplitude of AE signals in the anodic half-cycle, but, for the reason indicated above, the AE signal remains in the exposure cathodic half-cycle.

Therefore, the AE method makes it possible to detail the stage-by-stage course of the PEO process on various alloys surfaces and, by this way, to monitor in situ the evolution of the oxidation layer. It should also be emphasized that the information already available is sufficient to adjust the operating mode of the power supply or to temporarily turn off its individual circuits/processed objects.

4. Conclusions

The applicability of AE method for in-situ monitoring and optimization of PEO modes was demonstrated by the found exact correlation of acoustic emission modes with the stage-by-stage course of the PEO process established by other research methods. Identification and correction of PEOmode deviations from the standard (reference) mode, detailing the parameters of PEO process on each of the treated surfaces during group processing, and also the timely stop of PEO before the process begins to move over to the next (undesirable) stage, like to the area of destructive arc discharges - all this only makes an incomplete list of tasks that can be solved using AE-monitoring of PEO process today. This calls for establishing the reference values of AE signals in the amplitude and spectral regions that are most sensitive to the PEO stages, which is the subject of further research.

Acknowledgments

The research was financed by a grant from the Russian Science Foundation (Project No. 20-79-10262). The authors are grateful to Prof. Dmitry Merson (Togliatti State University) for recommendations on acoustic emission tests. The authors are also deeply grateful to Professor Mikhail Krishtal and Dr. Alexander Katsman (Togliatti State University) for their support.

References

[1] Hanshan D 2010 *Surface engineering of light alloys. Aluminium. magnesium and titanium alloys* (Woodhead Publishing Limited)

[2] Suminov I V 2011 *Plasma-electrolytic modification of the surface of metals and alloys* (Moscow: Technosphera)

[3] Boinet M, Verdier S, Maximovitch S et al. 2004 *NDT&E International* **37** 213–9

[4] Yerokhin A, Nie X, Leyland A et al. 1999 *Surface and Coatings Technology* **122** 73–93

[5] Clyne T, Troughton S 2019 *International Materials Reviews* **64(3)** 127–62

[6] Mukaeva V, Gorbatkov M, Parfenov E et al. 2018 *Electrotechnical and information complexes and systems* **14(3)** 60–5
[7] Ness S, Sherlock C, Moore P and McIntire P 1996 *American Society for Nondestructive Testing Nondestructive testing handbook (2nd ed.). Nondestructive testing overview* **10** 596 p
[8] Lazarev S, Mozgovoi A, Vinogradov A et al. 2009 *Journal of Acoustic Emission* **27** 212–23
[9] Rakoch A, Gladkova A, Linn Z and Strekalina D 2015 *Surface and Coatings Technology* **269** 138–44

Features of anodic plasma electrolytic treatment of an ultra-light alloy of the Mg-Li system in an aqueous electrolyte

I V Tambovskiy[1,2], B L Krit[1,3], S A Kusmanov[2], A S Perkov[2], N V Morozova[1,4], V M Medvetskova[3], R A Vdovichenko[1] and I R Palenov[1]

[1] Moscow State University of Technology "STANKIN", 127994, Moscow, Russia
[2] Kostroma State University, 156005, Kostroma, Russia
[3] Moscow Aviation Institute (National Research University), 125993, Moscow, Russia
[4] Russian Medical Academy of Continuing Professional Education, 123993, Moscow, Russia

E-mail: ramstobiliti@gmail.com

Abstract. The possibility of plasma electrolytic treatment of an alloy of the Mg-Li system in an aqueous solution of ammonium chloride, ammonium nitrate and boric acid has been studied. The specific features of the process in the voltage range from 10 to 600 V are determined. The current-voltage and voltage-temperature characteristics during alloy processing are shown, as well as the effect of the processing time on the temperature of the samples. Plasma electrolytic boronitriding of the alloy can be carried out to study the efficiency of the technology at a voltage of 240 V, which corresponds to a sample temperature of 300 ± 10 °C.

1. Introduction

Among the metal materials used in various branches of modern industry, in recent years, increased interest is shown in ultra-light magnesium alloys, which is due to their special properties in combination with low density. Materials based on Mg are particularly attractive for biomedical applications, since they are non-toxic, biologically and mechanically compatible with bone and muscle tissues. Magnesium is the fourth most abundant cation in the human body, is involved in many metabolic processes, and is the most biotolerant of metals [1]. For this reason, magnesium alloys are actively used for the manufacture of orthopedic, surgical and vascular implants, invasive devices, various instruments and parts of medical equipment. The use of ultra-light magnesium alloys can open up wide new opportunities for the development of more convenient devices with improved performance. At the same time, the main negative feature that significantly limits the spread of magnesium-based alloys is their high chemical activity and, as a consequence, low resistance to corrosion destruction in various environments. Tribological characteristics also have a very low level, which ultimately reduces the adaptive potential [2].

The solution to these problems can be the modification of the surface of magnesium alloys. For these purposes, plasma electrolytic treatment technologies are increasingly being used [3]. These technologies have found greater application in the processing of steels, cast irons and titanium alloys, contributing to the hardening of the surface layer due to the diffusion of atoms of light elements and hardening. This is further reflected in an increase in corrosion and wear resistance. Depending on the conditions of surface modification, it is possible to carry out a cathodic or anodic version of plasma

Content from this work may be used under the terms of the Creative Commons Attribution 3.0 licence. Any further distribution of this work must maintain attribution to the author(s) and the title of the work, journal citation and DOI.
Published under licence by IOP Publishing Ltd

electrolytic saturation of metals and alloys, which can be combined with plasma electrolytic polishing, as well as carrying out microarc oxidation.

Microarc oxidation is most often used to modify the surface of magnesium and its alloys. For example, microarc oxidation effectively protects a magnesium substrate from corrosion in an SBF solution, reduces the corrosion current density by more than 2 orders of magnitude, and the corrosion potential shifts in a positive direction [4].

However, the possibilities of plasma electrolytic saturation of Mg alloys have not been studied at all. At the same time, it is known that magnesium nitride is an effective catalyst for the structurization of solid refractory compounds [5], and the alloy of boron, magnesium and aluminum is known for its low coefficient of sliding friction and high hardness. By analogy with the role of the nitride phases for other valve metals [6], the positive effect of the magnesium alloy nitriding on its properties is not excluded.

In this sense, it is of interest to study the possibility of carrying out and features plasma electrolytic saturation of the Mg-8Li-1Al-0.6Ce-0.3Y alloy in an electrolyte containing boron and nitrogen.

2. Materials and methods

Plates of alloy Mg-Li (Mg-8Li-1Al-0.6Ce-0.3Y) 18 mm long, 5 mm wide and 2.5 mm high were treated in an electrolyte containing 10 % ammonium chloride, 5 % ammonium nitrate and 3 % boric acid using a cylindrical electrolyzer under conditions of longitudinal flow of the sample by the electrolyte. In the upper part of the chamber the electrolyte was being overflowed over the edge into the pallet, from where it was then pumped through the heat exchanger at flow rate of 2.5 L/min. The temperature of electrolyte cooled with tap water was maintained equal to (20±2) °C at the inlet to the electrolyser. During processing, the sample was connected to the positive pole of the DC power supply, and the working chamber to the negative one. After applying voltage to the sample, it was slowly lowered into the electrolyzer at a speed of 1 mm / s to a depth of 14 mm. To study the current-voltage and voltage-temperature characteristics, a voltage from 10 to 600 V was applied to the sample. The current intensity was measured using a DP6 10-2000A ammeter, and the process voltage was controlled using a DC DP6 2/20/200/600V DC voltmeter

The temperature of the sample was measured by a thermocouple made of chromel and alumel wire 0.5 mm in diameter, which were insulated, and at the end twisted and welded in graphite powder with borax at a potential of 200 V, set using an alternating current source. The thermocouple was connected to an APPA 103N digital multimeter and inserted into a 1 mm diameter hole in the plate, which was drilled with a drill to a depth of 5 mm. The upper part of the plate with the thermocouple was insulated with a heat-shrinkable tube to eliminate contact with the electrolyte and the purity of the experiment.

3. Results and discussion

Figure 1 show the current-voltage characteristic obtained during the anodic plasma electrolytic treatment (PET) of the alloy. Depending on the voltage applied to the sample, the nature of the process changes. In the voltage range of 10–30 V, an electrolysis regime is observed (T_{sample} = 80–110 °C (figure 2)), which changes to an interrupt regime when the voltage increases to 40 V and exists up to 90 V inclusively (T_{sample} = 60–95 °C (figure 2)). In the interrupt regime, the current values vary in the range of 1.5–8.5 A (figure 1). In these conditions, the energy is insufficient for the formation of plasma around the anode, however, a high speed of anodic dissolution of the alloy is observed.

After increasing the voltage to 100 V, an electrolytic plasma is formed and the interrupt regime changes to the heating regime (T_{sample} = 200 °C (figure 2)). When the voltage increases to 440 V, the plasma resistance rises and the current intensity reduces (figure 1), as also the temperature of the sample increases (figure 2). At the starting voltage U_0 = 240–320 V, reported to the system, and the contact of the sample with the electrolyte, the alloy heats up to extremely high temperatures and reacts chemically with water, which is accompanied by burning and the formation of the corresponding oxides. So the existence of stationary heating is valid at U_0 = 120–220 V, at which a dielectric oxide layer forms on the surface of the alloy in contact with the electrolyte, preventing heating to a critical temperature and burning in the aqueous medium.

At U_0 = 350–400 V and operating U = 320–400 V, there is a plasma electrolytic polishing regime in which, depending on the voltage, the sample temperature does not exceed 120 °C (figure 2)). In these conditions, the current is minimal and increases with increasing voltage (figure 1). Reducing the working voltage to $U \leq 310$ V leads to a change in the polishing regime to the heating regime and the alloy is heated again to a critical temperature and burns in the aqueous medium.

At higher voltages $U \geq 460$ V, the anodic polishing regime changes to the electrolyte spraying regime, which indicates the maximum plasma resistance and is accompanied by a decrease in the current intensity (figure 1). In these conditions, the sample temperature does not exceed 100 °C (figure 2).

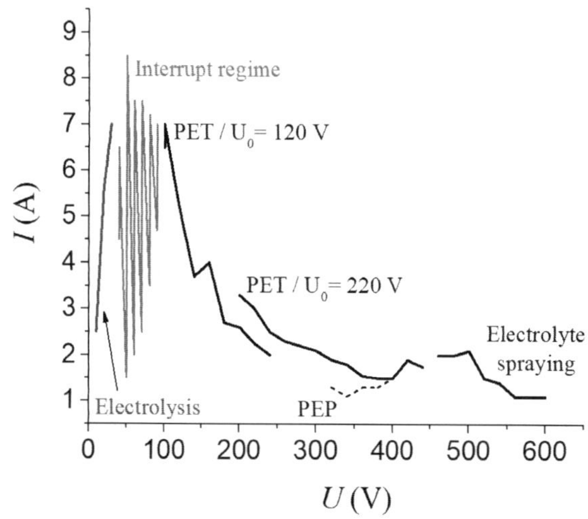

Figure 1. Current-voltage characteristic during plasma electrolytic treatment of Mg-LI alloy in the electrolyte containing NH_4Cl, NH_4NO_3 and H_3BO_3.

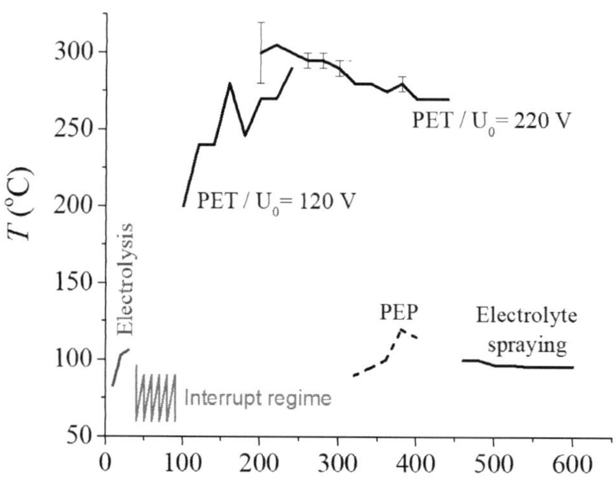

Figure 2. Voltage-temperature characteristic during plasma electrolytic treatment of Mg-LI alloy in the electrolyte containing NH_4Cl, NH_4NO_3 and H_3BO_3.

Figure 3 shows the dependence of the temperature of samples from the Mg-Li alloy on the time of plasma electrolytic treatment at constant voltage. It was found that the T_{sample} = 300±5 °C does not change for 5 minutes at a processing voltage of 240 V. That is why PET can be carried out in these conditions, for example, by varying the process time.

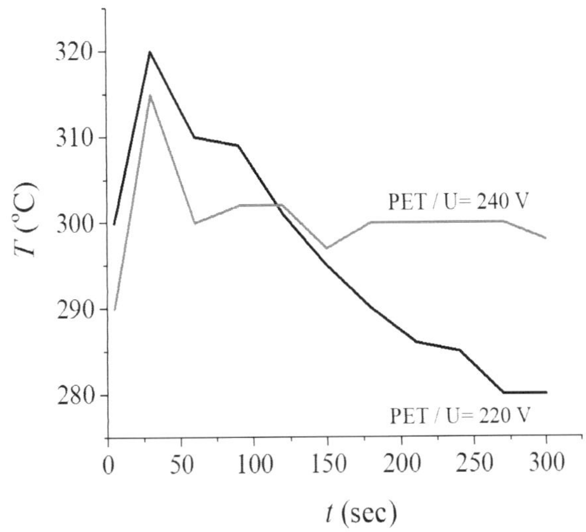

Figure 3. The dependence of the temperature of samples from the Mg-Li alloy on the time of plasma electrolytic treatment at constant voltage in the electrolyte containing NH_4Cl, NH_4NO_3 и H_3BO_3.

4. Conclusions

The possibility of plasma electrolytic treatment of an alloy of the Mg-Li system in an aqueous solution of ammonium chloride, ammonium nitrate and boric acid has been studied.

It was found that during processing in the voltage range from 10 to 30 V the electrolysis regime is observed, from 40 to 90 V – the interruption regime, from 100 to 440 V – the heating regime, from 460 to 600 V – the electrolyte spraying regime.

At the starting voltage U_0 = 240–320 V, reported to the system, and the contact of the sample with the electrolyte, the alloy heats up to extremely high temperatures and reacts chemically with water, which is accompanied by burning and the formation of the corresponding oxides.

At U_0= 350–400 V and operating U = 320–400 V, there is a plasma electrolytic polishing regime.

Acknowledgments

This work was financially supported by the Russian Science Foundation (Contract No. 21-79-30058). During the work, the equipment of the Center for Collective Use of Moscow State University of Technology "STANKIN" was used.

References

[1] Shilov A M, Melnik M V and Osia A O 2008 *Rus. Med. J.* **21** 1439–44

[2] Wu R, Yan Y, Wang G, Murr L E, Han W, Zhang Z and Zhang M 2015 *Inter. Mater. Rev.* **2** 65–100

[3] Belkin P N, Borisov A M and Kusmanov S A 2016 *J. Surf. Invest.* **10** 516–35

[4] Zhao L, Cui Ch, Wang Q et al. 2010 *Corr. Sci.* **7** 2228–34

[5] Wentorf Jr R H 1961 *J. Chem. Phys.* **34(3)** 809–12

[6] Sobol O V, Andreev A A, Grigoriev S N, Volosova M A and Gorban V F 2012 *Met. Sci. Heat Treat.* **54** 28–33

Anomalous evolution of the ion-induced surface relief of highly oriented pyrolytic graphite

N N Andrianova[1], A M Borisov[1,2], E S Mashkova[3], M A Ovchinnikov[3] and V S Sevostyanova[2]

[1] Moscow Aviation Institute (National Research University), Volokolamskoe sh. 4, 125993 Moscow, Russia
[2] Moscow State University of Technology "STANKIN", Vadkovsky lane 1, 127055 Moscow, Russia
[3] Skobeltsyn Institute of Nuclear Physics, Moscow State University, Leninskie gory 1(2), 119991, Moscow, Russia

E-mail: anatoly_borisov @mail.ru

Abstract. The modification of the surface of highly oriented pyrolytic graphite (HOPG) under 10, 20 and 30 keV Ar^+ ions irradiation with fluence 10^{18} cm^{-2} at the irradiation temperature of 250°C has been studied experimentally. An anomalous growth of the ion-induced surface relief of HOPG have been found. This effect, like the well-known effect of anomalous deep embedded argon ions in HOPG, is analyzed within the framework of plastic deformation mechanisms in graphite.

1. Introduction

Understanding and accounting for the modification of material properties under ion irradiation are necessary for many ion-plasma technologies and are especially relevant for carbon materials because of their widespread usage in radiation fields of different nature, including plasmas of hydrogen and helium isotopes in fusion plants [1–3]. Along with the nanostructural features of ion-induced surface structures, such as etching pits, cones, and ripples [4], specific macroscopic changes in the topography of irradiated surfaces are observed with increasing irradiation fluences of carbon materials. These are, for example, corrugations on carbon fibers [5, 6], columnar-needle morphological elements [7–9], and elements with a macro-scale "ruck&tuck" geometry on highly oriented pyrolytic graphite (HOPG) [10]. The formation of such structures is attributed to the fact that graphite, like other two-dimensional layered materials, can deform by the bending, kinking, and twinning as a result of the combined motion of basal dislocations and layers sliding [10–12]. It was also found that high-fluence irradiation of HOPG with argon ions with energies of 10-30 keV resulted in an anomalously deep modification of the surface [7–9]. In this work, we investigated the evolution of the ion-induced topography of the HOPG surface under conditions of the maximum manifestation of this effect.

2. Experimental

As the targets in the experiment, we used samples of HOPG grade UPV-1T in the form of plates with a basal plane (001). Preparation of the target prior to irradiation included obtaining a close to a mirror surface by removing the upper layers of the target with a scotch tape. Irradiations were carried out on a

Content from this work may be used under the terms of the Creative Commons Attribution 3.0 licence. Any further distribution of this work must maintain attribution to the author(s) and the title of the work, journal citation and DOI.
Published under licence by IOP Publishing Ltd

mass-monochromator of the Skobeltsyn Institute of Nuclear Physics, Moscow State University [13] by Ar^+ ions with energies of 10 - 30 keV along the normal to the surface of samples. Irradiation fluence was ~10^{18} cm^{-2}. The irradiation temperature was varied from room temperature to 400°C. The target surface topography was studied using a VEGA 3 LMH scanning electron microscope (TESCAN, Brno, Czech Republic).

3. Results and discussion

SEM images in figure 1 show the topography of HOPG at an irradiation temperature of 250°C, which corresponds to the maximum effect of anomalous deep surface modification [7–9]. The top row on figure 1 shows the side view of SEM images with the tilt angle to the surface of 30°. The bottom row shows the top view of SEM images obtained at normal incidence of the probing electron beam.

When the ion energy E = 10 keV, not exceeding the energy threshold of deep modification (E_{th} = 15 keV [8]), the characteristic elements of side view relief are single cones with density more than 10 cones/μm^2 and height h about 0.2 μm. At anomalously deep modification of the surface layer at ion energy $E > E_{th}$, when the embedded argon is found in the layer of depth h_{Ar} by orders of magnitude greater than the projective range R_p of ions in graphite, the surface is dominated by submicron columnar-needle morphological elements of an order of magnitude greater (1-2 μm) height compared to single cones at $E < E_{th}$. The top view of all SEM images shows a random fibrillar nanostructure with numerous bends and kinks in the fibrils.

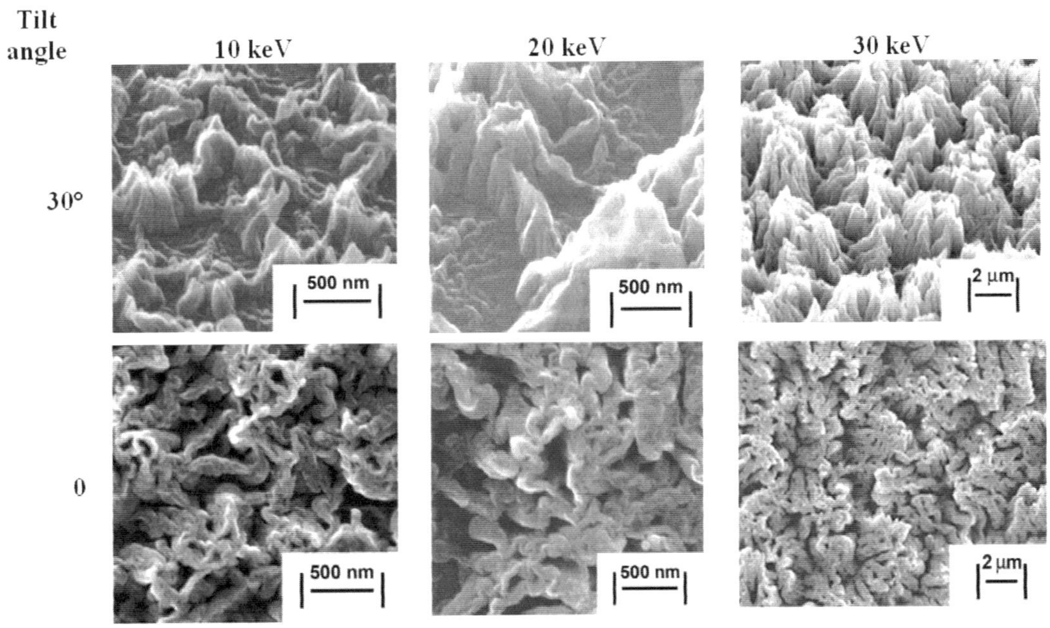

Figure 1. SEM images of HOPG with a sample inclination of 30° and along the normal to the surface after irradiation with 10, 20 and 30 keV Ar^+ ions at the temperature of 250°C.

The side view of SEM images was used to estimate the height of the ion-induced morphological elements h at various ion energies (figure 2). It is seen that the height h significantly increases with the ion energy and is 4-5 times greater than the depth of argon h_{Ar} embedded into the surface layer.

The evolution of the ion-induced HOPG topography with increasing ion energy expands the view of anomalous deep modification, which in [8–9] has been associated with the intense diffusion of the interstitial carbon atoms along the inclined to the basal plane graphene planes of the ion-induced twins, leading to the formation of an altered layer with implanted ions at depths $h_{Ar} >> R_p$. The transition of the ion-induced HOPG topography from cones at $E < E_{th}$ to a submicron columnar-needle topography at an anomalously deep surface modification with $h_{Ar} >> R_p$ suggests significant mass transfer in the HOPG surface layer at $E > E_{th}$, which may occur by twinning, bending, kinking, and folding plastic deformations and their mixed types typical for two-dimensional layered materials [8–11]. The corresponding deformation patterns can be seen in the top view of the SEM images (bottom row in figure 1). At the ion energy of 10 and 20 keV ions, the nanostructure is mostly characterized by bends around the normal to the surface. At the ion energy of 30 keV, the bends of carbon fibrils perpendicular to the surface plane become prominent in the nanostructure, which leads to the overlapping and layering of some fibrils over others.

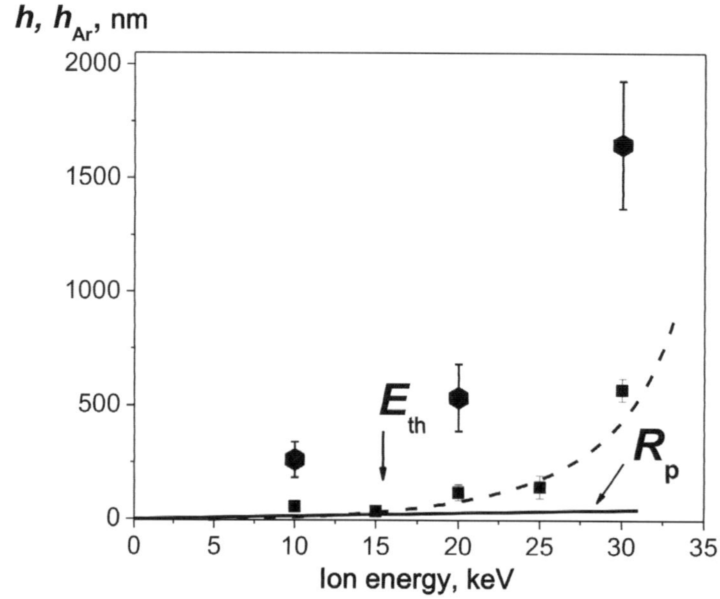

Figure 2. Dependences of the height of morphological elements h on the surface of HOPG (hexagons) and the depth of argon embedding h_{Ar} [7] (squares) on the energy of Ar^+ ions. E_{th} - threshold energy of the anomalously deep modification effect.

The energy threshold for the deep modification effect may be related to the thresholds for the activation of various mechanisms of deformation of two-dimensional materials over their thickness [12]. In this work it have been found that when two-dimensional crystal thin films, which have high strength properties without breaking even at significant angles of mechanical bending, are mechanically stressed, the resulting tensions are mostlly relaxed through plastic deformation processes. Thus, the angle of mechanical bending and the number of basal layers in the film are the main parameters influencing the type of plastic deformation process. If the crystal is thin, around 30 layers and thinner, bending deformations dominate. If a crystal is thick, the nonbasal twinning occurs along with bends.

Under ion irradiation, the mechanical stresses are caused by gradient ν(x) of the radiation damage profiles in the number of *dpa* [5–11]. The profiles ν(x) corresponding to the present experiment are shown in figure 3. Calculations were carried out according to [5, 6]. The *dpa* number $\nu = \Phi \cdot \sigma_{dam}$, where Φ – irradiation fluence, σ_{dam} – the cross section of radiation damage, is the useful value to compare the damage effect under different irradiation conditions. To evaluate ν we used a computer simulation of the ion interaction with solids using the SRIM code [14]. Estimations were performed for a flat graphite surface at normal ion incidence. The simulation results in the form of the profile of the average number of displacements formed by one ion per path length $\sigma_{dam}(x) \cdot n_0$, where n_0 – atomic density, is used to calculate the stationary profile $\nu_{st}(x)$ at high fluence in the dynamically equilibrium irradiation conditions at $\Phi > R_d \cdot n_0 / Y$,

$$\nu_{st}(x) = \frac{n_0}{Y}\int_x^{\infty} \sigma_{dam}(x') \cdot dx' \qquad (1)$$

where Y is the sputtering yield, R_d - the depth of defect production. The depth of defect formation R_d determining the upper limit of the integration was taken to be equal to the depth, which in the $\sigma_{dam}(x) \cdot n_0$·profiles accounts for 99% of the deposited damage.

Figure 3 shows that the profile depth d increases proportionally with the ion energy. Estimates show that at E_{th} = 15 keV, the depth d is of the same order of magnitude of the number of layers in the ion-modified layer as the number of layers in a two-dimensional crystal thin film when the transition from bending to twinning occur [12]. Thus, with an increase in the ion energy, the depth of defect formation increases and, as a consequence, the number of layers involved in deformation processes increases.

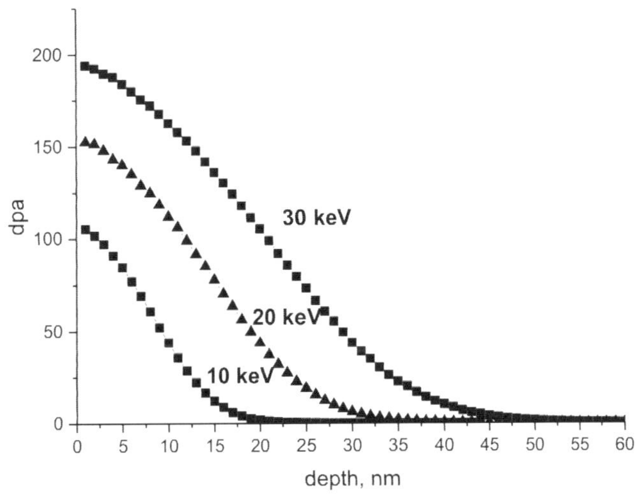

Figure 3. DPA profiles for graphite irradiated by Ar^+ ions with energies of 10, 20, and 30 keV.

This leads to the predominance of nonbasal twinning in the modified layer. An increase in the number of layers deformed by twinning leads to an almost complete absence of flat surface areas corresponding to the initial HOPG surface (cf. the side view of the SEM image at 10 and 30 keV in figure 1). Under extended irradiation, this will lead to sliding and deformation processes in directions with a significant inclination to the initial basal plane, as a result of which, the embedded argon will be at considerable depths, compared to the projectile range.

4. Conclusions

The modification of the surface of highly oriented pyrolytic graphite under 10, 20 and 30 keV argon ions irradiation with irradiation fluence 10^{18} cm^{-2} at a temperature of 250°C have been studied experimentally. An anomalous growth of the ion-induced surface relief of HOPG with a threshold at an ion energy of 15 keV have been found. It is shown that this effect, as well as the previously found effect of anomalous deep embedded argon ions in highly oriented pyrolytic graphite, is associated with significant changes in the surface geometry as a result of graphite-specific ion-induced plastic deformation processes.

Acknowledgments
This research was carried out with the financial support of the Russian Science Foundation within the framework of scientific project No. 21-79-30058.

References

[1] Virgil'ev Yu S and Kalyagina I P 2004 *Inorganic Materials* **40** S33–S49
[2] Behrisch R 2010 *J Synch. Investig.* **4** 549–62
[3] Grigoriev S, Peretyagin P, Smirnov A, Solis W, Diaz L A, Fernandez A and Torrecillas R 2017 *J. Eur. Ceram. Soc.* **37** 2473–79
[4] Carter G 2001 *J. Phys. D: Appl. Phys.* **34** R1–22
[5] Andrianova N N, Borisov A M, Mashkova E S, Parilis E S and Virgiliev Yu S 2013 *Ion-Induced Modification and Crimping of Carbon Composite Fibers* (Horizons in World Physics vol 280) ed A Reimer (New-York: Nova Science Publishing) chapter 5 pp 171–90
[6] Andrianova N N, Borisov A M, Vysotina E A, Timofeev M A, Mashkova E S and Ovchinnikov M A 2021 *J. Synch. Investig.* **15** 230–5
[7] Andrianova N N, Avilkina V S, Borisov A M and Mashkova E S 2012 *Nucl. Instrum. Methods in Phys. Res. B.* **273** 58–60
[8] Andrianova N N, Borisov A M, Mashkova E S, Sevostyanova V S and Virgiliev Yu S 2013 *Nucl. Instrum. Methods in Phys.Res. B.* **315** 117–20
[9] Andrianova N N, Borisov A M, Virgiliev Yu S, Mashkova E S and Sevostyanova V S 2014 *Bull. Russ. Acad. Sci. Phys.* **78(6)** 520–5
[10] Bacon D J and Rao A S 1980 *Journal of Nuclear Materials* **91** 178–88
[11] Liu D, Cherns D, Johns S, Zhou Y, Liu J, Chen W-Y, Griffiths I, Karthik C, Li M, Kuball M, Kane J and Windes W 2021 *Carbon* **173** 215–31
[12] Rooney A P, Li Z, Zhao W, Gholinia A, Kozikov A, Auton G, Ding F, Gorbachev R V, Young R J and Haigh S J 2018 *Nature Communications* **9** 3597
[13] Mashkova E S and Molchanov V A 1985 *Medium-Energy Ion Reflection from Solids.* (Amsterdam: North-Holland)
[14] Ziegler J F and Biersack J P 2013 SRIM http://www.srim.org.

Influence of argon ion irradiation dose on the formation of the surface layers composition and changing mechanical properties of ion-plasma coated carbon steel

P V Bykov[1], V L Vorob'ev[1], S G Bystrov[1], V V Tarasov[1], A Yu Drozdov[1], V Ya Bayankin[1] and S S Alexandrova[2]

[1] Udmurt Federal Research Center, Ural Branch, Russian Academy of Sciences, 426067, Izhevsk, Russia
[2] Moscow Aviation Institute (National Research University), 125993, Moscow, Russia

E-mail: bykovpv@udman.ru

Abstract. The effect of argon ion irradiation with an energy of 40 keV in the dose range of 10^{16} - 10^{18} ion/cm^2 on the formation of the surface layers composition, changes in the morphology and mechanical properties (microhardness, and wear resistance) of carbon steel AISI 1020 with a deposited ion-plasma coating $Ni_{80}Cr_{20}$ was studied. It is shown that irradiation with doses greater than 10^{17} ion/cm^2 leads to the formation of a layer consisting of nickel, chromium and iron. The most optimal treatment mode for improving wear resistance is irradiation with a dose of $5 \cdot 10^{17}$ ion/cm^2.

1. Introduction

During friction, the fracture process is localized in the surface layers of the material. Therefore, with the advent of new technologies that affect the surface, the problem arose of their effective use for surface hardening of machine parts [1, 2]. These methods include various coating processes. Combining the coating process with ion implantation can increase the thickness of the hardened layer and its physical, mechanical and operational properties [3–7].

The aim of this work was to study the effect of argon ion irradiation at various doses on the formation of the composition and structure of surface layers, changes in the morphology and mechanical properties (microhardness and wear resistance) of carbon steel AISI 1020 with deposited ion-plasma coating $Ni_{80}Cr_{20}$.

2. Samples and research methods

Samples of carbon steel AISI 1020 were plates with a length of 30 mm and cross section of 8 × 2 mm. The composition of the initial samples was as follows: Fe, base; C, 0.2%; Mn, 0.4%; and Si, 0.15%. The samples were annealed at a temperature of 650 ^{0}C for 30 min in a vacuum of ~10^{-4} Pa. The deposition of an ion-plasma coating $Ni_{80}Cr_{20}$ with a thickness of 100 nm and irradiation with argon ions with various doses of 10^{16}, $5 \cdot 10^{16}$, 10^{17}, $5 \cdot 10^{17}$ and 10^{18} ion/cm^2 was carried out on the "Yashma-4" installation. The ion energy was 40 keV and the ion current density was 278 μA/cm^2.

Topography studies were carried out by atomic force microscopy (AFM) on a SOLVER 47 PRO probe microscope in contact mode. The arithmetic mean surface roughness of the R_a samples under

Content from this work may be used under the terms of the Creative Commons Attribution 3.0 licence. Any further distribution of this work must maintain attribution to the author(s) and the title of the work, journal citation and DOI.
Published under licence by IOP Publishing Ltd

study was calculated for each sample from AFM images of 15 1x1 micron surface areas using a probe microscope data processing program.

The chemical composition of the surface layers was studied by X-ray photoelectron spectroscopy (XPS) on a SPECS spectrometer using AlKα-radiation (1486.6 eV) in combination with layer-by-layer etching of the surface with argon ions. The calculated rate of etching was 1 nm/min. The phase composition of the samples was studied on a Rigaku Ultima IV multifunctional diffractometer using CuKα radiation.

The microhardness of the surface layers of the samples before and after irradiation was measured by pressing a diamond indenter on a PMT-3M device at a load of 10 gram. To increase the reliability of the obtained result, the measurement procedure was performed 10 times or more. Tribological tests were carried out on an SRV-III friction machine in the reverse friction mode with a relative reciprocating motion of the contact pair. Test scheme: polished ball Ø 10 mm (steel AISI 52100) - plane. The normal load is 20 N. The oscillation frequency is 2 Hz, the amplitude is 2 mm. The test time is 10 minutes. The temperature is 24...27 °C.

3. Results and discussion

Atomic force microscopy studies indicate changes in surface morphology as a result of irradiation with argon ions (figure 1). On the surface of the samples irradiated with argon with doses of 10^{16} и $5\cdot10^{16}$ ion/cm^2, bands oriented along the samples are observed. Perhaps this is due to the fact that the deposited $Ni_{80}Cr_{20}$ coating repeats the traces of polishing on the carbon steel AISI 1020 surface. With an increase in the irradiation dose to 10^{17}, $5\cdot10^{17}$ ion/cm^2, separate peaks appear on the surface. After irradiation with argon ions with a dose of 10^{18} ions/cm^2, structures similar to blisters appear on the surface [8]. The arithmetic mean roughness of the surface R_a, depending on the irradiation dose with argon ions, behaves non-monotonically and does not correlate with the maximum height difference (table 1).

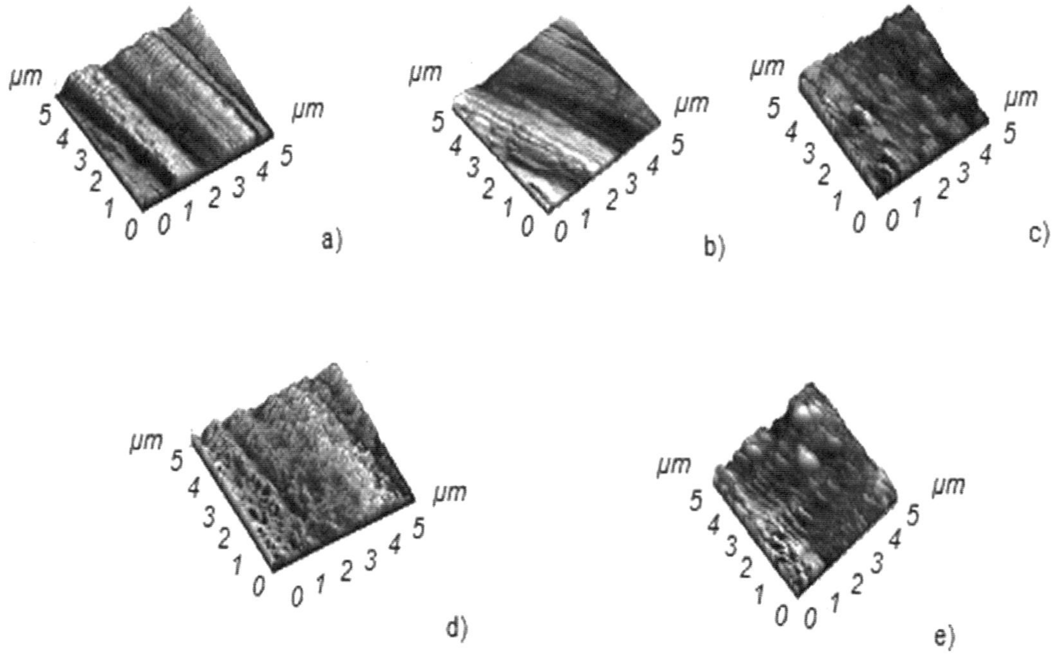

Figure 1. Topography of samples surface after argon ion irradiation with a dose of: a) 10^{16}, b) $5\cdot10^{16}$, c) 10^{17}, d) $5\cdot10^{17}$, e) 10^{18} ion/cm^2.

Table 1. Roughness of samples after irradiation.

Radiation dose	Roughness R_a, nm	Mean square deviation, nm	Maximum height difference, nm
10^{16} ion/cm^2	8,1	2,0 (18%)	60
$5\cdot10^{16}$ ion/cm^2	6,9	1,6 (23%)	110
10^{17} ion/cm^2	6,5	1,3 (20%)	100
$5\cdot10^{17}$ ion/cm^2	11,1	2,0 (18%)	120
10^{18} ion/cm^2	6,9	1,7 (25%)	100

The results of studying the composition of the surface layers of carbon steel AISI 1020 samples with deposited ion-plasma coating $Ni_{80}Cr_{20}$ after argon ion irradiation are shown in Figure 2. Irradiation with a dose of 10^{16} – $5\cdot10^{16}$ ion/cm^2 does not lead to the penetration of nickel and chromium into thin surface layers of steel. This can be explained by the fact that the projective range of argon ions in $Ni_{80}Cr_{20}$ is much less than the thickness of the deposited layer. The calculation by the SRIM method [9] showed that the projective range of argon ions with an energy of 40 keV in $Ni_{80}Cr_{20}$ is $R_p \approx 17.2$ nm, and the longitudinal deviation $\Delta R_p \approx 10$ nm. At the same time, a large amount of carbon and oxygen is observed in the surface layers (up to 30 nm). Nickel and chromium in the deposited layer are in a pure state. With an increase in the irradiation dose to 10^{17} - 10^{18} ion/cm^2, the thickness of the $Ni_{80}Cr_{20}$ coating decreases due to sputtering [8] and a layer consisting of nickel, chromium and iron is formed. The XPS spectra show peaks corresponding to compounds of iron and chromium with carbon.

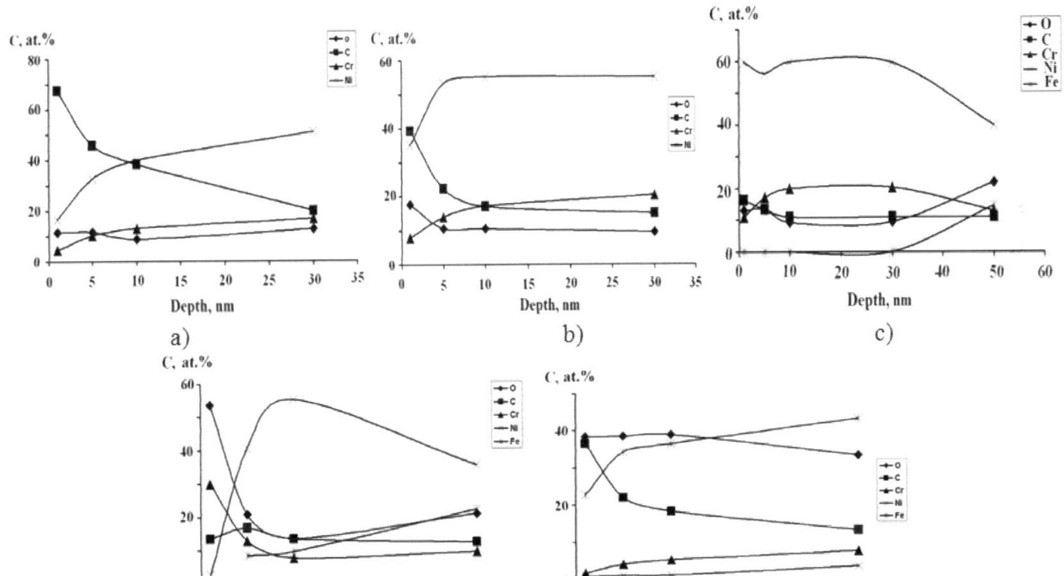

Figure 2. Concentration profiles of element depth distribution depending on the argon ion irradiation dose: a) 10^{16}, b) $5\cdot10^{16}$, c) 10^{17}, d) $5\cdot10^{17}$, e) 10^{18} ion/cm^2.

The study of the phase composition of the coated samples showed that the surface of all samples, regardless of the radiation dose, is two-phase and consists of α- and γ-phases. A quantitative analysis of the phase content showed that with an increase in the irradiation dose, the proportion of the γ-phase decreases, while the lattice period of the α-phase increases (table 2).

The deposition of an ion-plasma coating and subsequent irradiation with argon ions leads to a decrease in the microhardness (table 2) of the surface layers from 1000 MPa, in the initial state, to 850 MPa (10^{16} and 5×10^{16} ion/cm^2) and 770 MPa (10^{17}, 5×10^{17} and 10^{18} ion/cm^2) after irradiation.

Table 2. Volume fraction of γ-phase, lattice period of α-phase and microhardness of samples after irradiation.

Radiation dose	Volume fraction of γ-phase, %	a, nm	Microhardness, MPa
10^{16} ion/cm^2	23±3	0,2864±0,0001	841±52
$5\cdot10^{16}$ ion/cm^2	24±3	0,2865±0,0001	842±55
10^{17} ion/cm^2	20±3	0,2866±0,0001	769±44
$5\cdot10^{17}$ ion/cm^2	13±2	0,2866±0,0001	769±46
10^{18} ion/cm^2	8±3	0,2867±0,0001	774±48

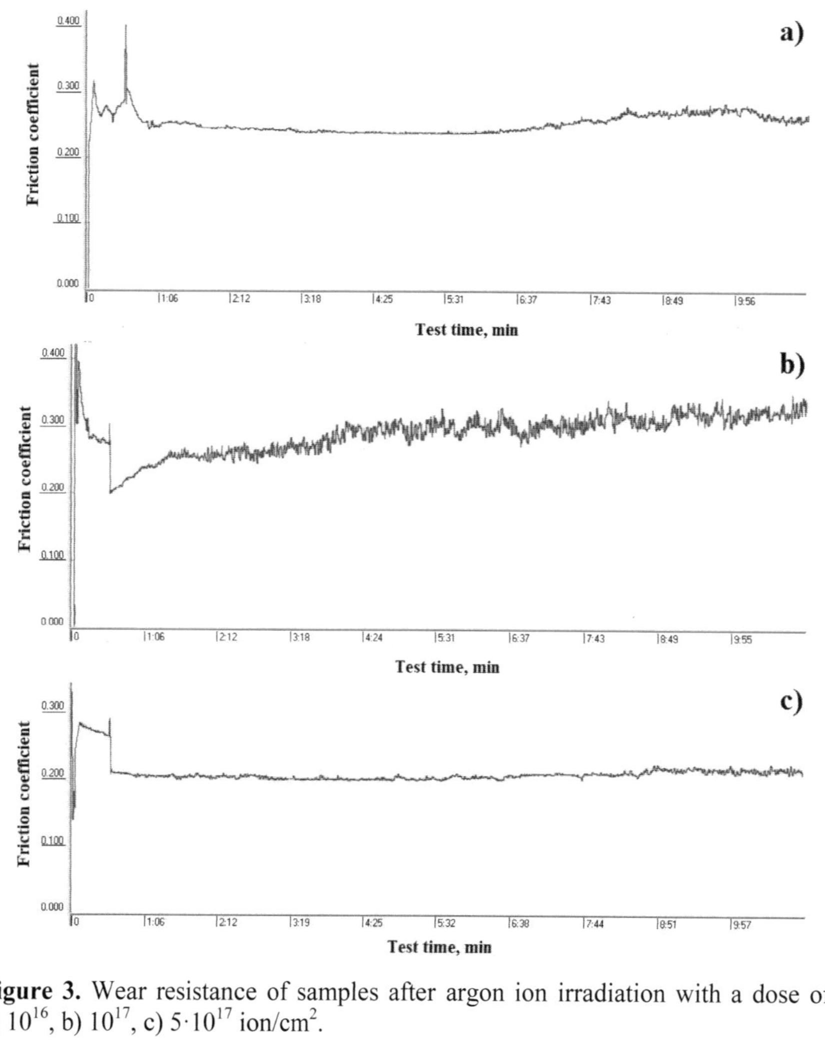

Figure 3. Wear resistance of samples after argon ion irradiation with a dose of: a) 10^{16}, b) 10^{17}, c) $5\cdot10^{17}$ ion/cm^2.

Tests for wear resistance have shown that the coefficient of friction of carbon steel AISI 1020 samples in the initial state (without coating and ion implantation) is 0.27. The deposition of the $Ni_{80}Cr_{20}$ coating and the subsequent implantation of argon ions leads to a change in the friction coefficient (Figure 3), the best result was achieved on a sample irradiated with a dose of $5 \cdot 10^{17}$ ion/cm^2 – 0.2. For the rest of the samples, changes in the coefficient of friction were recorded in the range from 0.23 to 0.32 (on average). Samples irradiated with a dose of 10^{16} and $5 \cdot 10^{17}$ ion/cm^2 showed the most stable results in terms of the amplitude of fluctuations in the coefficient of friction. Almost throughout the entire test cycle, the value of the friction coefficient does not change, which indicates the high quality of the modified layer. Obviously, during the tests, the treated surface is not subject to destruction, which affects the nature of the frictional interaction of the friction pair. A similar result is achieved on a sample irradiated with a dose of 10^{18} ions/cm^2, the amplitude of the friction coefficient fluctuations remains unchanged and rather low during the tests. The most unstable results are observed on a sample irradiated with a dose of 10^{17} ion/cm^2, for which the amplitude of oscillations varies within wide limits. It is likely that during testing, already at the initial stage (running-in stage), a partial destruction of the treated surface occurs, and then this process is aggravated. The performance of the pair is also affected by the roughness of both contacting surfaces. In this case, the roughness of the steel ball is noticeably higher than the roughness of a flat sample with a modified surface, which can significantly distort the test results [1].

4. Conclusions

The deposition of an ion-plasma coating $Ni_{80}Cr_{20}$ on the surface of carbon steel AISI 1020, and subsequent argon ion irradiation with doses of more than 10^{17} ion/cm^2 leads to the formation of a layer consisting of nickel, chromium and iron. The most optimal treatment mode for improving wear resistance is irradiation with a dose of $5 \cdot 10^{17}$ ion/cm^2.

Acknowledgments

The work was carried out within the State Assignment of the Ministry of Science and Higher Education of the Russian Federation No. 121030100002-0.

This study was performed using equipment of the Shared Use Centre "Centre of Physical and Physicochemical Methods of Analysis and Study of the Properties and Surface Characteristics of Nanostructures, Materials, and Products" UdmFRC UB RAS.

The authors of the article express their gratitude to Professor V.P. Krivobokov (Tomsk Polytechnic University) for his help in preparing samples.

References

[1] Tushinsky L I 1990 *Metall alloys strengthening theory and technology* (Novosibirsk: Nauka. Sibirian Division) 306 p
[2] Budzynski P 2015 *Nucl. Instrum. Methods Phys. Res. Sect. B* **342** 1–6
[3] Kostyuk G I 2007 *The effective cutting tools having the coating and hardened layers* (Kharkov: National aerospace university named by N E Gukovsky «Kharkov aviation institute») 633 p
[4] Pogrebnyak A D, Martynenko V A, Mikhalev A D, Shablya V T and Yanovskii V P 2001 *Tech. Phys. Let.* **27** 615–7
[5] Kalin B A, Volkov N V and Oleinikov I V 2012 *Bull. Russ. Acad. Sci.: Phys.* **76** 690–5
[6] Bocharov G S, Eletskii A V, Zakharenkov A V, Zilova O S, Sliva A P, Terentyev E V, Fedorovich S D and Churilov G N 2018 *J. Surf. Invest.: X-ray, Synchrotron Neutron Tech.* **12** 27–32
[7] Vorob'ev V L, Gilmutdinov F Z, Syugaev A V, Bykov P V and Bayankin V Ya 2020 *Phys. Met. Metallogr.* **121** 460–5
[8] Behrische R 1981 *Sputtering by Particle Bombardment* (Heidelberg: Springer-Verlag) p 336
[9] Ziegler J F SRIM-2013, http://www.srim.org

Mechanical properties and fatigue of extruded D16 alloy profiles of the Al-Cu-Mg system after ion irradiation

N V Gushchina[1], V V Ovchinnikov[1,2], K V Shalomov[1] and D I Vichuzhanin[2]

[1] Institute of Electrophysics, Ural Branch of Russian Academy of Sciences, Amundsena Street 106, Yekaterinburg, 620016, Russia
[2] Institute of Engineering Science, Ural Branch of Russian Academy of Sciences, Komsomolskaya Street 34, Yekaterinburg, 620049, Russia

E-mail: guscha@rambler.ru, guschina@iep.uran.ru

Abstract. The effect of 20-keV Ar^+ ion irradiation on the mechanical properties and fatigue resistance of 6 mm thick hot-extruded profiles made of a D16 alloy (Al–Cu–Mg system) has been studied after natural aging. Irradiation at fluences of $2{\cdot}10^{15}$ cm^{-2} and $1{\cdot}10^{16}$ cm^{-2} has been revealed to increase the number of cycles to failure significantly at low load amplitudes σ/σ_u = 0.3. The maximum increase in the number of cycles to destruction 2.4 times is observed at a smaller fluence of ions $F = 2{\cdot}10^{15}$ cm^{-2}.

1. Introduction

This work is to investigate the effect of Ar^+ ion irradiation on the fatigue resistance of hot-extruded D16 alloy (Al–Cu–Mg system) profiles in the as-delivered state (after quenching and natural aging).

The D16 alloy, the classic duralumin, is characterized by high strength and high ductility. It is widely used as a structural material in various fields of modern technology, including aircraft construction [1, 2]. The changes in the microstructure, phase composition and mechanical properties of cold-worked VD1 and D16 alloy sheets (Al–Cu–Mg system) 2–3 mm thick after Ar^+ ion irradiation were previously presented in [3]. The ion-beam treatment regimes were selected to ensure the softening of these alloys to the required level of properties, which allows further cold-rolling of the alloys. Based on these results, we patented the technology of intermediate radiation annealing of commercial aluminum alloys during their cold rolling, which provides a 2–3 times reduction of labor and energy consumption of the processes as compared to heat treatment [3].

Therefore, it is of interest to search for opportunities to enhance the mechanical properties and resource characteristics of the D16 alloy using ion-beam irradiation as a finishing treatment.

2. Experimental part

Hot-extruded PR-100-23 profiles (6 mm thick) [4] made of the D16 alloy after quenching and natural ageing were provided by Kamensk-Ural'skii Metallurgical Works. Standard samples for mechanical and fatigue tests were previously made from these profiles.

The samples were irradiated using an ILM-1 ion beam implanter equipped with a PULSAR-1M ion source based on a glow discharge with a hollow cold cathode [5]. A ribbon ion beam 100×20 mm^2 was cut from the cylindrical ion beam using a collimator. To decrease the heating temperature, the samples were irradiated from both sides while moving them under the ion beam at a speed of 2.5 cm/s. The

Content from this work may be used under the terms of the Creative Commons Attribution 3.0 licence. Any further distribution of this work must maintain attribution to the author(s) and the title of the work, journal citation and DOI.
Published under licence by IOP Publishing Ltd

temperature of the samples during irradiation was controlled with a chromel–alumel thermocouple welded to a similar test sample and a computerized digital signal measurement system based on ADAM-4000 modules. The maximum temperature to which the samples were heated during irradiation was below 30–40°C. We used the following ion beam parameters: ion energy E = 20 keV, ion current density j = 200 μA/cm^2, and fluences $F_1 = 2\cdot10^{15}$ cm^{-2} (irradiation time was 1.6 s) and $F_2 = 1\cdot10^{16}$ cm^{-2} (irradiation time was 8 s).

Samples were tested on an INSTRON-8801 universal servohydraulic tester equipped with a computer and a FastTrack controller. All operations during the experiment were performed using hardware-software tools. Sinusoidal cycle fatigue tests were performed at a loading frequency of 3 Hz to determine the Weller fatigue curve. The asymmetry coefficient of the cycle was –1.

3. Details of exposure, measurements, and results

As a result of tests for uniaxial tension of the initial and irradiated samples of D16 alloy, it was found that after quenching and natural aging, D16 alloy has the following mechanical properties: $\sigma_в$ = 557 МПа, $\sigma_{0,2}$ = 441 МПа, δ = 9,7 %. The mechanical properties after ion irradiation at fluence $F = 2\cdot10^{15}$ cm^{-2} remain almost the same. Irradiation at a higher fluence $F = 1\cdot10^{16}$ cm^{-2} increases the relative elongation to δ = 12.3 %, without changing the strength properties (changes within the measurement error): σ_u = 545 MPa and $\sigma_{0.2}$ = 435 MPa.

Figure 1 shows the Weller curves built based on the results of the tests of the initial and irradiated D16 alloy samples. The number of cycles to destruction at σ/σ_u = 0.3 is seen to increase significantly after irradiation under the used conditions. The number of cycles to destruction for the initial sample is 633 757 cycles, whereas it is 1 523 825 cycles after irradiation at a fluence of $2\cdot10^{15}$ cm^{-2} and 1 000 000 cycles at a fluence of $1\cdot10^{16}$ cm^{-2}.

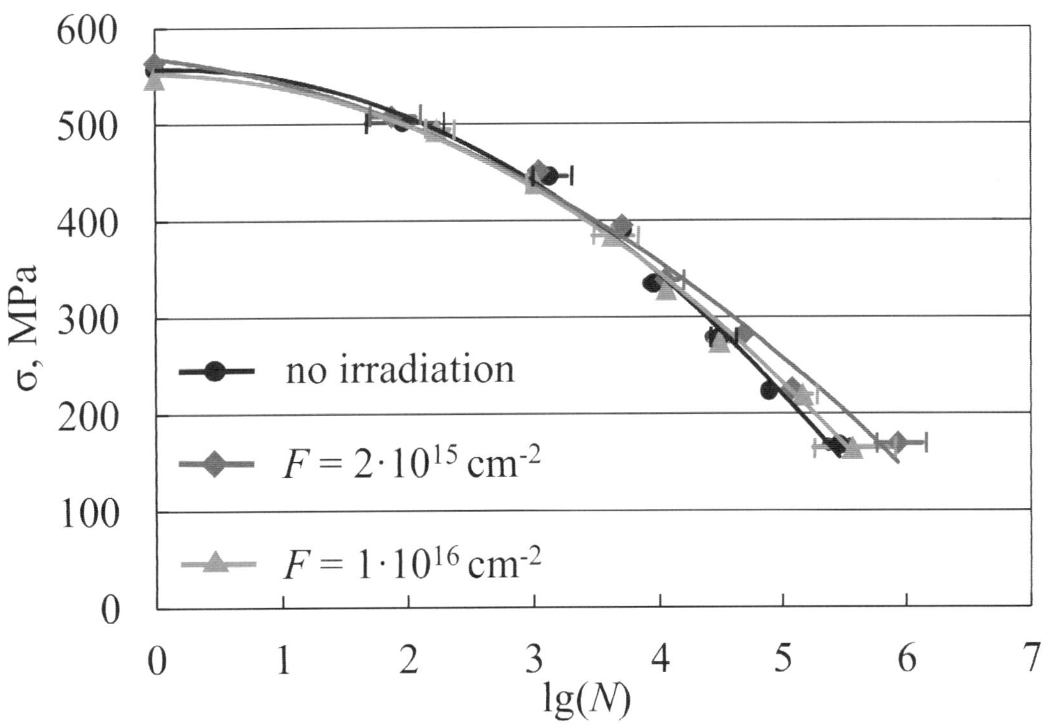

Figure 1. Number of cycles lg(N) as a function of maximum cyclic stress σ for the D16 alloy samples before and after Ar+ ion irradiation (E = 20 keV).

This phenomenon may be due to the fact that small elastic deformations occur under conditions of multi-cycle fatigue [6, 7]. Moreover, the surface and the volume of the irradiated sample are subjected to instant (for several seconds) ion-beam radiation annealing [3]. As a result, both the surface and the volume become almost free of deformation defects. Therefore, the failure starts from the sample surface much later, since the processes involving damage accumulation and cracking develop much slower. Low-cycle fatigue causes elastoplastic deformations and defects accumulate across the entire sample cross section [6, 7].

4. Conclusions

This research showed that short-term (8 s) Ar^+ ion irradiation of the D16 alloy at an ion energy of 20 keV and fluence $F = 1{\cdot}10^{16}$ cm^{-2} increased the plasticity of the alloy and retained its strength characteristics. The irradiation at fluences of $2{\cdot}10^{15}$ cm^{-2} and $1{\cdot}10^{16}$ cm^{-2} was revealed to increase the number of cycles to failure significantly at low load amplitudes $\sigma/\sigma_u = 0.3$. The maximum 2.4-time increase in the number of cycles to destruction was observed at a smaller fluence of ions $F = 2{\cdot}10^{15}$ cm^{-2}.

Acknowledgments

The investigations were carried out using the equipment of the Plastometry Testing Center for Collective Use at the Institute of Machine Science, Ural Branch, Russian Academy of Sciences.

This work was supported in part by RFBR grant № 19-08-00802-a.

References

[1] Kolachev B A, Elagin V I and Livanov V A 2001 *Metal Science and Thermal Treatment of Nonferrous Metals and Alloys* (Moscow: MISiS)

[2] Friedlander I N 1979 *Aluminum deformable constructive alloys* (Moscow: Metallurgy)

[3] Ovchinnikov V V 2018 *Surface and Coating Technology* **355** 65–83

[4] GOST 13737-90 1990 *Extruded rectangular profiles with equal angular section from aluminum and magnesium alloys. Range* (Moscow: Publishing house of standards)

[5] Gavrilov N V, Mesyats G A, Nikulin S P, Radkovskii G V, Eklind A and Perry A J 1996 *J. Vac. Sci. Technol.* A **14** 1050–6

[6] Terentyev V F and Oxogoev A A 2001 *Cyclic strength of metallic materials* (Novosibirsk: NSTU Publishing House)

[7] Nicholas T 2006 *High Cycle Fatigue. A Mehanics of Materials Perspective* (Oxford UK: Elsevier Ltd)

Arc-PVD coatings for titanium machining tool

I V Blinkov, A P Demirov, D S Belov and A V Chernogor

NUST «MISiS», 125993, Moscow, Leninskiy prospect 4, Russia

E-mail: biv@misis.ru

Abstract. The paper presents results of life tests in machining titanium alloy with cutting carbide tool with Ti-Mo-Al-Ni-Si-N, (Ti, Al) N coatings and without it. Shown, that tool life with first coatings is 1.8 and 3.6 times higher than tool with second coatings and simple carbide tool, respectively.

1. Introduction

The creation of new structural materials with high hardness and heat resistance requires the development of new technological processes of machining materials using high cutting speeds and the rejection of coolant. The desire to use a coating with the same composition and structure in few machining operations (milling, turning, plaining and shaping) requires steady improvement of the multifunctional coatings applied to the cutting tool. The reliability of coated cutting tools can be increased by reducing the physicochemical interaction of the processed material and the coating. In this case, it is necessary that the coating has high values of hardness and fracture toughness, increased tribological properties and high adhesion to the substrate [1–7].

In our previous research [8] we discussed deposition process of Mo-Si-Al-Ti-Ni-N coating on nickel-based alloy and its properties. The coating is characterized by a multilayer architecture with a crystallite size of about 6–10 nm. It had high hardness, adhesive strength with the substrate, and a low coefficient of friction.

In the paper considered the structure of Ti-Mo-Al-Ni-Si-N coating deposed by arc-PVD on carbide cutting tool, investigated its mechanical properties, and compared toll life with and without cathodic coatings. The results obtained will make it possible to determine promising areas of application of Ti-Mo-Al-Ni-Si-N coatings.

2. Materials and methods

Ti-Mo-Al-Ni-Si-N coatings were obtained by arc-PVD installation equipped with three cathodic systems with toroidal electromagnetic droplet phase separators. The cathodes were made of molybdenum, Al-Si (20 at. %), and Ti-Ni (50 at.%) alloys. $(Ti_{0.7}, Al_{0.3})N$ coatings used as control sample was deposed the same method with using Ti-Al(50 at. %) cathode. The current of evaporating arc (I_a) was 120 A. A negative base voltage supplied on carbide alloy substrate was equal – 120 V. Deposition was in reaction gas atmosphere (N_2). Pressure of nitrogen was approximately $3 \cdot 10^{-1}$ Pa. Deposition process duration for each coating was 90 minutes. The temperature substrate was 400 °C. The thickness of coatings was from 3.9 to 4.1 µm. The cross-sectional image of coatings is shown in figure 1. As shown in the figure, Ti-Mo-Al-Ni-Si-N coating is characterized by a multilayer architecture, while $(Ti_{0.7}, Al_{0.3})$ N coating is columnar-structured.

Content from this work may be used under the terms of the Creative Commons Attribution 3.0 licence. Any further distribution of this work must maintain attribution to the author(s) and the title of the work, journal citation and DOI.
Published under licence by IOP Publishing Ltd

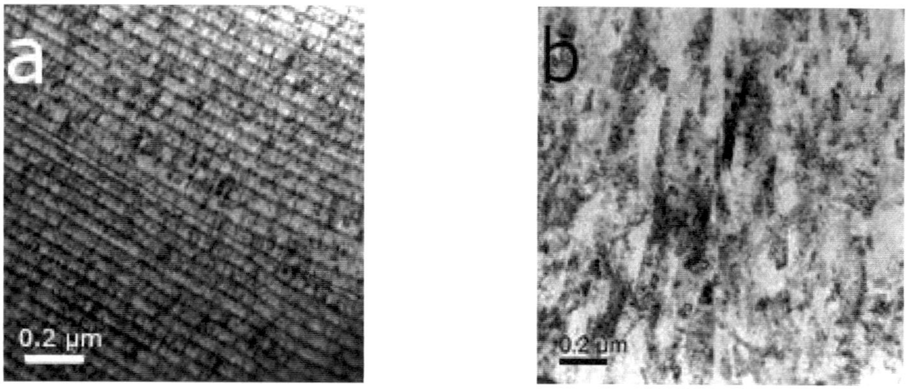

Figure 1. TEM-images of Ti-Mo-Al-Ni-Si-N **(a)** and Ti- Al- N **(b)** coatings.

Life-tool testing with single-point turning tools was carried out according to ISO 3685:1993. Machining material (workpiece) was titanium alloy cylindrical billet with a diameter of 200 mm. The alloy composition was Ti – 86.45–90.9; Fe till 0.6; C till 0.1; Si till 0.1; V 3.5–5.3; Al 5.3–6.8; Zr till 0.3 wt. % (HB 10^{-1} = 255–341 MPa, E = 115 GPa). Testing carbide cutting tools has size 12,7×12,7×4,75 mm (SNUN 120408). The following cutting mode was implemented: cutting speed V = 150 m / min, feed S = 0.1 mm / rev, cutting depth t = 0.5 mm. The criterion for the wear of the cutting tool was the flank wear h^{max} = 0.5 mm. h was measured by optical microscope. Tool life and path, while h^{max} was achieved were obtained. The components of the cutting forces (F_x, F_y, F_z) were measured (figure 2).

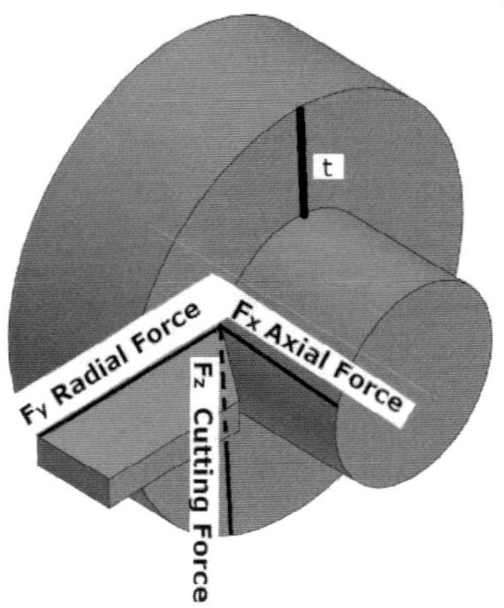

Figure 2. Schematic representation of the cutting forces in the life-tool test.

3. Results and discussion

Mechanical properties of the tested coatings are given in table 1. The Ti-Mo-Al-Ni-Si-N coating is characterized in relation to the reference sample by the best values of hardness, parameters that determine the resistance to elastic (H/E) and plastic deformation (H^3/E^2), characteristics of adhesion strength to the substrate and a lower value of the coefficient of friction (μ). Figure 3 shows the images of the flank face of uncoated and coated carbide tools after 7 minutes of the tool-life test (longitudinal turning). The main types of wear, in this case, are abrasion and adhesive interaction with the workpiece material. The developed coating slows down wear compared to Ti-Al-N coated and uncoated tools.

The determination of cutting characteristics (Fig. 4, 5) results show that the tool life with a Ti-Mo-Al-Ni-Si-N coating increases at 3.6 and 1.8 times in comparison with an uncoated tool and coated ($Ti_{0.7}$, $Al_{0.3}$)N, respectively.

Table 1. Physical and mechanical characteristics of coatings on WC-Co alloy substrate.

	H, GPa	*E, GPa*	W_p, %	H/E	H^3/E^2	μ	*σ, GPa*	L_{c1}, N	L_{c2}, N
($Ti_{0.7}$,$Al_{0.3}$)N	30.1±1,2	475±23	40	0.06	0.12	0.75	-2.1	5.1	31.2
Ti-Mo-Al-Ni-Si-N	37.6±1,1	547±27	38	0.07	0.18	0.60	-2.3	11.9	53.5

W_p – plastic deformation.
σ – residual stress determined by the Stoney formula.
L_{c1} – the critical failure load.
L_{c2} – full delamination of the stack.

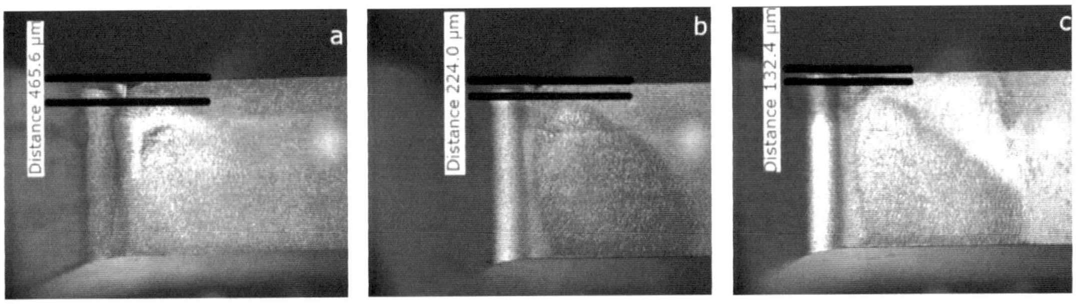

Figure 3. Images of the worn flank of carbide tool at 7 minutes tests during continuous turning of titanium alloy without (a) and with Ti-Al-N (b) and Ti-Mo-Al-Ni-Si-N (c) coatings.

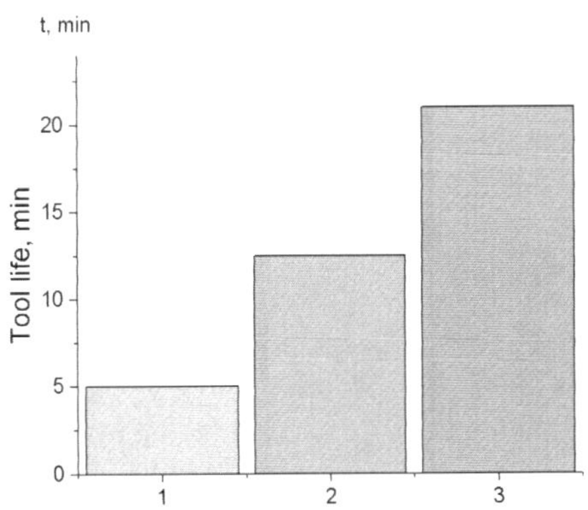

Figure 4. Tool life when turning titanium alloy: (1) uncoated; (2) with (Ti, Al) N and (3) Mo-Si-Al-Ti-Ni-N coatings.

At the same time, the use of Ti-Mo-Al-Ni-Si-N coatings leads to a decrease in cutting forces by up to 2 and 1.5 times in comparison with the reference samples. The results obtained are associated with a lower friction coefficient of Ti-Mo-Al-Ni-Si-N coatings, their multilayer architecture, increased adhesion strength to the base and hardness.

4. Conclusions

Durability cutting tool testing when turning titanium alloy shows that Ti-Al-Ni-Si-N coatings increase lifetime tool in 3.6 and 1.8 compared with uncoated and (Ti, Al)N coated, respectively. At the same time, the use of Ti-Mo-Al-Ni-Si-N as hardening coatings leads to a decrease in cutting forces by up to 2 and 1.5 times in comparison with the reference samples. The lifetime difference is determined by the increased physical and mechanical properties of the Ti-Mo-Al-Ni-Si-N coating, its tribological characteristics, and multilayered architecture.

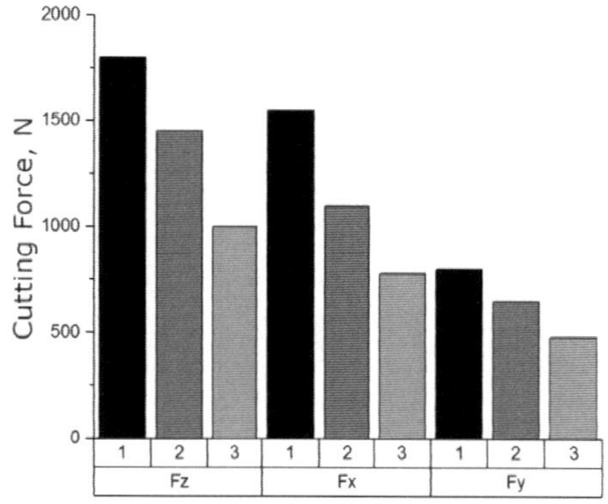

Figure 5. Influence of coatings on the components of the cutting force when turning a titanium alloy with a carbide cutting tool (1 - uncoated tool; 2 - with (Ti, Al) N coating; 3 - with Mo-Si-Al-Ti-Ni- N coating)

Acknowledgments
This study was supported by the Russian Scientific Foundation, project no. 19-19-00555.

References

[1] Vereschaka A S, Grigoriev S, Sitnikov N et al. 2017 *Wear* **390** 209–19
[2] Yin-Yu Chang and Hsing-Ming Lai 2014 *Surface & Coatings Technology* **259** 152–8
[3] Beake B and Fox-Rabinovich G 2014 *Surface & Coatings Technology* **255** 102–11
[4] Krishnaraj V, Samsudeensadham S, Sindhumathi R et al. 2014 *Procedia Eng.* **97** 251–7
[5] Bozakis K-D, Michailidis N, Skordaris G et al. 2012 *CIRP Annals* **61** 703–23
[6] Vereshchaka A A, Vereshchaka A S, Mgaloblishvili O et al. 2014 *JAMT* **72** 303–17
[7] Krella A 2020 *Coatings* **10** 921
[8] Blinkov I V, Volkhonskii A O, Chernogor A V et al. 2018 *Inorganic Materials* **5** 437–45

The applying of layer-by-layer laser remelting in the technology of laser metal deposition

Y N Zavalov and A V Dubrov

Institute on Laser and Information Technologies - Branch of the Federal Scientific Research Centre "Crystallography and Photonics" of Russian academy of Sciences, Svyatoozerskaya 1, 140700, Shatura, Moscow Region, Russia

E-mail: dubrov.av@mail.ru

Abstract. Layer-by-layer laser remelting (LRM) was used in the technology of laser metal deposition in order to reduce the porosity of the layer in the case of single-layer and multi-layer coatings. The layer porosity is shown to decreases with LRM at a rate 5 times higher than the rate of metal deposition. The dependence of the porosity estimated using the method of the hydrostatic weighing on the rate of laser remelting is obtained.

1. Introduction

Reduction of porosity in additive technologies of laser deposition of metals (LMD) is the urgent task of the modern stage of equipment introduction into modern machine-building production of complex shaped products [1–2]. Porosity also occurs in SLM [3] and is associated with the features of the technological process [4], in particular, the occurrence of convective flow in the melt bath [5]. There was indicated in [6] the need for additional drying of the powder in preparation for production. The dependence of porosity on technological parameters was studied in [7] for LMD of titanium aluminium alloys, and stainless steels was used in [2]. In this work, laser remelting technology is used on each layer in the LMD process in the case of single-layer and multi-layer coatings. Earlier, the technology of laser remelting (LRM) a single track formed using the LMD technology was studied in [8, 9]. The mode of melting the track with the incomplete depth was studied in [8], but no data on porosity in the LMD + LRM mode was given. The combined technology was studied in addition to the single track in [9] for the formation of the thin-walled structure. The porosity was shown to decrease first with the increase the power of laser radiation for the single track, and it begins to increase then from the certain power level. The laser power of the optimal LRM technology was concluded to coincide with the power at LMD. The laser power remained on certain level, and the energy effect of laser radiation varied due to changes in the scanning speed in our experiments.

2. Materials and method

The experiments were carried out at the additive synthesis research facility using the technology of laser deposition of metals, created at IPLIT RAS. The flow of PR-07X18H12M2 steel powder (analogue of 316L) with the mass flow rate of 0.14 mg/s and the granulometric composition (40 ... 100) microns was concentrated through the conical nozzle on the surface of the substrate. 68x68x6 mm substrates made of 08X18H10T steel were used. The fibre ytterbium laser LK-400-V was used, the radiation of which passes through the coaxial flow of the gas-powder mixture and focuses near the

Content from this work may be used under the terms of the Creative Commons Attribution 3.0 licence. Any further distribution of this work must maintain attribution to the author(s) and the title of the work, journal citation and DOI.
Published under licence by IOP Publishing Ltd

substrate. Each track with the length of 16 mm was formed at constant values of the distance from the nozzle to the substrate and the position of the beam focus, scanning speed *V*=350 mm/min, laser power *P*=400 W.

Tracks with *dy* increments were applied sequentially to each substrate, depending on the degree of overlap. The height of the formed single layer is indicated depending on the degree of overlap. The height change was taken into account in the case of a subsequent remelting operation, see Table1. The images of the formed layers before remelting (a) and after (b) are shown on figure 1. The seven adjacent tracks were formed sequentially in one direction for a small overlap in total, and 8 tracks were formed for 40% overlap.

Table 1. The changing the height of the layer after remelting.

Overlap	Speed *V*	Height *H* [LMD]	Height *h* [LMD+ LRM]	(*H-h*)	Deviation
%	mm/min	microns	microns	mm	%
10%	350	730	600	130	18%
25%	350	980	880	100	10%
40%	350	1188	1100	88	7%

The value of the vertical step *dz* depended on the degree of overlap in the case of multilayer coatings. More than 12 layers were applied to form the rectangular part with the height of about 16 mm (figure 1 c). The gas pressure (nitrogen) at the feeder inlet was 0.3 MPa, the pressure of the protective gas was 0.6 MPa, the flow rate of the protective gas was 10 nl/min. Samples were made with different degrees of overlap of the tracks in the layer. Samples were made with a layer-by-layer LRM operation in the combined technology [LMD+LRM], the geometry reproduced the geometry of sample manufacturing, but the vertical step value *dz* was adjusted in this case, table 1.

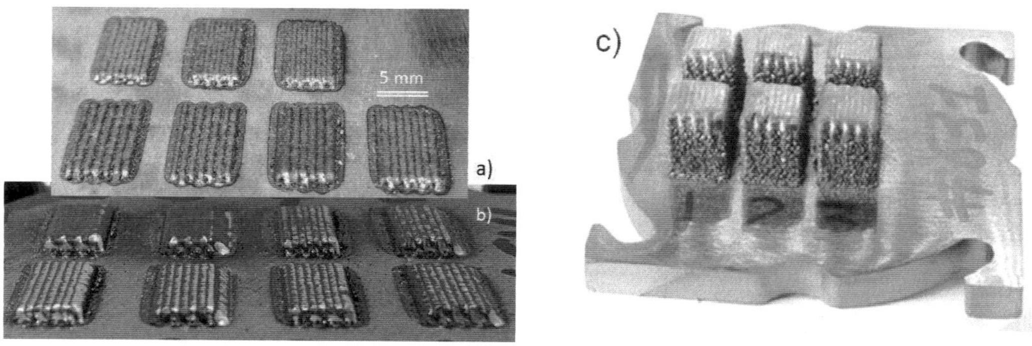

Figure 1. The part of the obtained specimens: a) and b) two-layer and c) multi-layer specimens.

Then the samples underwent the operation of removing the top layer, and the porosity of each sample was measured by hydrostatic weighing ("Archimedes ' method"). The weighing instrument was used with the minimum division value of 0.1 mg. The obtained results are shown in figure 2. The samples are divided by the degree of overlap and are listed at the same degree in ascending order of the LRM speed in the figure. The empty markers indicate the speed of the layered LRM operation, the last one shows the porosity of the sample without the LRM operation (a large circle). The increase in the LRM rate, the effect of reducing porosity worsens in general. The first two samples were obtained at LRM rate significantly lower than the sample formation rate for each overlap value. The porosity at a reduced LRM rate significantly worsened compared to the [LMD+LRM] technology in the case of overlap degrees of 10% and 40%. The porosity worsened was not observed in the case of an overlap degree of 25%. |And special mention the initially porosity was noticeably less in samples with the overlap degree of 25%.

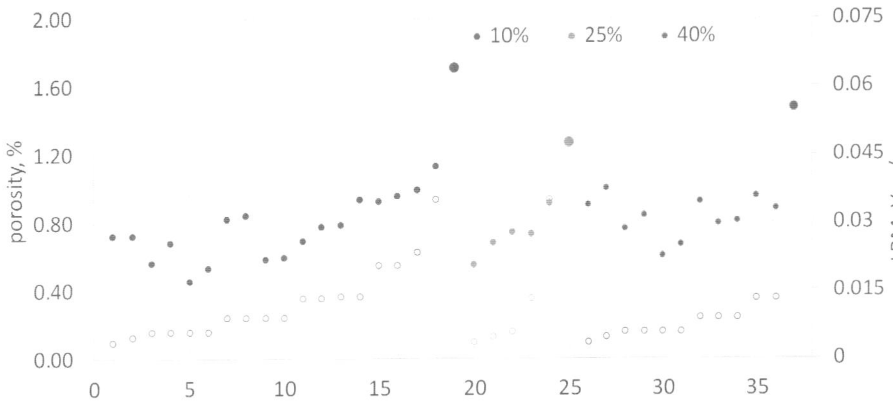

Figure 2. The dependence of the porosity value for different LM+LP samples. The degree of overlap of the tracks in the layer is indicated in the figure caption. Hollow markers - the scanning speed of the LP (right scale).

3. Results and discussion

To estimate the thermal effect of laser radiation on metal, we use the value of the energy *Es* per unit surface:

$$E_S = \frac{P}{2\,\omega\cdot V} \tag{1}$$

To estimate the thermal effect of laser radiation on metal, we use the value of the energy *Es* per unit surface area, where ω is the radius of the laser beam in the plane. Previously, the study of the residual intra-layer porosity (ILP) in multilayer samples obtained by the isolated LMD technology was carried out in the energy supply range (1.2...9) kJ/cm^2 [10]. The dependence of the ILP on the specific energy is shown non-monotonic, the minimum of defects is observed at the average values of the energy from the used range. Figure 3 shows the dependence of the porosity of the obtained data and Figure 2 shows the dependence on the specific energy of irradiation *Es* at the layer-by-layer LMR stage. The samples are formed into three groups depending on the degree of overlap, and in each group, they are numbered in ascending order *Es*. The effect of reducing porosity worsens at above 6 kJ/cm^2 with an overlap of 10%, at the degree of overlap of 25% the porosity also decreased at an *Es* value of about 10 kJ/cm^2.

4. Conclusions

The parametric study of the residual porosity of multilayer LMD+ LRM samples obtained at different speeds of the LRM process and different degrees of overlap of the tracks in the layer is carried out.

The addition of the LRM stage with the selected parameters is shown to make it possible to achieve the decrease in porosity in all the studied LMD modes. The minimum porosity values achieved after LRM are from 0.43% to 0.95%. That is about 2-3 times less than the initial porosity of the LMD. When using LMR speeds of 5...6 times higher than the LMD speed, a porosity reduction of about two times is achieved. The influence of the degree of overlap of the tracks in the layer on the porosity is shown. The optimal overlap value corresponds according to the minimum porosity criterion to the degree of overlap closer to 25%. The porosity decreases with an increase in the specific irradiation energy *Es* at the layer-by-layer LMR stage. However, the porosity increases with the increase in *Es* above 6 kJ/cm^2 in the case of an overlap degree of 10%.

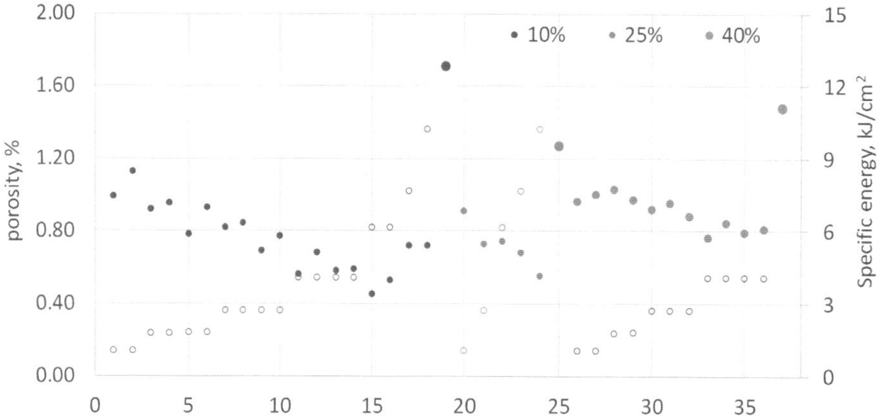

Figure 3. The dependence of the porosity value for different LM+LP samples. The degree of overlap of the tracks in the layer is indicated in the figure caption. Hollow markers – the dependence of the specific irradiation energy Es at the layer-by-layer LMR stage (right scale).

Acknowledgements
The work was carried out with the support of the Ministry of Science and Higher Education of the Russian Federation within the framework of the State Task of the FSRC "Crystallography and Photonics" of the Russian Academy of Sciences in terms of setting the problem, as well as with the grant support of the Russian Foundation for Basic Research in terms of the developed methodology, as well as in terms of the experimental results obtained, grant No. 20-21-00158.

References

[1] Li L 2006 *J. Mater. Sci.* **41** 7886–93
[2] Susan D F, Puskar J D, Brooks J A and Robino C V 2006 *Mater. Charact.* **57** 36–43
[3] Grasso M and Colosimo B M 2017 *Meas. Sci. and Technol.* **28** 044005
[4] Kobryn P A, Moore E H and Semiatin S L 2000 *Scr. Mater.* **43** 299–305
[5] Ng G K L, Jarfors A E W, Bi G and Zheng H Y 2009 *Appl. Phys. A* **97** 641
[6] Zhong C, Gasser A, Schopphoven T and Poprawe R 2015 *Opt. & Laser Technol.* **75** 87–92
[7] Khanzadeh M, Chowdhury S, Marufuzzaman M, Tschopp M A and Bian L 2018 *J. Manuf. Syst.* **47** 69–82
[8] Pôrto R M, de Souza Pinto Pereira A and Pereira M 2020 *J. of Laser Appl.* **32** 022069
[9] Xin B, Zhou X, Cheng G, Yao J and Gong Y 2020 *Opt. & Laser Technol.* **127** 106087
[10] Zavalov Y N and Dubrov A V 2021 *J. Phys. Conf. Ser.* (IOP Publishing) **1954** 012056

Cathodic boriding and anodic polishing of medium-carbon steel by plasma electrolysis

S S Korableva[1], I R Palenov[2], I M Naumov[1], A A Smirnov[1], I A Kusmanova[1], I S Gorokhov[1], R D Belov[1], E V Sokova[1] and K I Bestchetnikova[1]

[1] Department of Physical, Mathematical and Natural Science, Kostroma State University, 17 Dzerzhinskogo Street, Kostroma 156005, Russia
[2] Department of High-Efficiency Machining Technologies, Moscow State University of Technology "STANKIN", 3a Vadkovsky lane, Moscow 127994, Russia

E-mail: sakusmanov@yandex.ru

Abstract. The possibility of cathodic plasma electrolytic boriding of medium-carbon steel in an aqueous solution of ammonium chloride and boric acid followed by anodic plasma electrolytic polishing in an ammonium sulfate solution on the same equipment with a change in the operating voltage is shown. The morphology and roughness of the surface, microhardness of the modified layer have been investigated. Wear resistance was studied under dry friction conditions. It has been established that cathodic boriding at 850 °C for 5–30 min leads to the hardening of the surface layer up to 1050 HV with an increase in roughness by 1.5–2.5 times and wear resistance by 3.5 times. Subsequent anodic plasma electrolytic polishing of the boriding surface leads to a decrease in roughness with an increase in wear resistance by 2.3 times.

1. Introduction

The study and optimization of the modes of plasma electrolytic technologies for processing metal products to improve their performance, in particular, plasma electrolytic diffusion saturation, remain relevant in modern science, as evidenced by the regular publication of research results in this area [1–2].

The work [3] describes the experience of anodic plasma electrolytic boriding of medium-carbon steel in an electrolyte containing 10 % ammonium chloride and 5 % boric acid. Chemical-thermal saturation of the surface in the presence of boron in plasma electrolysis at 170 V for 5 minutes at 850 °C allows the formation of an outer oxide layer (FeO, Fe_3O_4) and a diffusion layer (Fe_2B, FeB) in the surface layer of steel, followed by martensite, formed by quenching at the end of the process. At the same time, as a result of anodic dissolution, the average surface roughness (up to 0.3 μm) of the part is reduced by more than 3 times compared to the untreated sample (1.0 μm). Anodic boriding leads to a significant increase in the microhardness in the modified layer (1800 HV) and increases the wear resistance of the alloy by 16 times.

Cathodic plasma electrolytic boriding (CPEB) of low-carbon and medium-carbon steels (4340, 4140, 1045, 3215 and 1020) in borax-based solutions provides the formation of borided layers with a microhardness of at least 900 HV in 10–15 min at 600 V [4]. The wear resistance of steel H13 enhances by 12 times after PEB in a solution containing borax and sodium hydroxide. The samples

Content from this work may be used under the terms of the Creative Commons Attribution 3.0 licence. Any further distribution of this work must maintain attribution to the author(s) and the title of the work, journal citation and DOI.
Published under licence by IOP Publishing Ltd

were tested for abrasive wear according to the "pin-on-disc" scheme against SiC at a normal load of 50 N at a constant rotation speed of 200 rpm [5]. However, the possibilities of CPEB in an electrolyte containing ammonium chloride and boric acid have not been studied. In this sense, it is of interest to study the effect of electrolyte components and the duration of cathodic boriding on the surface morphology, structural-phase composition in the surface layer of medium-carbon steel, microhardness distribution and surface wear resistance. It is known that during cathodic treatment on the sample surface, high-temperature oxidation and diffusion of saturating components always compete with the action of electric discharges [6], which usually prevail over oxidation. In some cases, under the action of discharges in electrolysis plasma, the outer oxide layer is deformed; craters are formed, which leads to the development of the relief and an increase in the roughness of the products. The solution to these problems can be a combination of cathodic plasma electrolytic saturation and anodic plasma electrolytic polishing (APEP).

2. Materials and methods

Cylindrical samples of medium-carbon steel (0.45 wt.% C) 11 mm in diameter and 15 mm in height were polished with sandpaper to the surface roughness of $R_a = 1.0 \pm 0.1$ μm and cleaned with acetone in ultrasonic bath. The CPEB was carried out at the sample temperature of 850 °C in an aqueous solution of 10 % ammonium chloride and 3 % boric acid using cylindrical electrolyser under conditions of longitudinal electrolyte flow around the sample. In the upper part of the chamber the electrolyte was being overflowed over the edge into the pallet, from where it was then pumped through the heat exchanger at flow rate of 2.5 L/min. The temperature of electrolyte cooled with tap water was maintained equal to (30 ± 2) °C at the inlet to the electrolyser. During processing, the sample was connected to the negative pole of the DC power supply, and the working chamber to the positive one. After applying a voltage of 220 V, the sample was slowly lowered into the electrolyser at a speed of 1 mm/s to a depth equal to its height. Then the voltage was reduced to 80 V that corresponded to a surface temperature of 850 °C. CPEB was carried out from 5 to 30 min, after which the samples were cooled in the electrolyte (quenched). Then samples were washed with water and dried.

APEP was carried out in an aqueous solution of 5 % ammonium sulfate at a voltage of 325 V. The depth of immersion of the upper end of the samples was 5 mm. A part of the anode current lead submerged in the electrolyte was insulated with a Teflon bushing to prevent the electrolyte from splashing and dissolving the current lead during APEP. The electrolyte was pumped through a heat exchanger at a flow rate of 1 l/min. The electrolyte temperature was maintained at 70±2 °C. The PEP time was 1 and 2 min.

The Micromed MET optical metallographic microscope with digital image visualization served to study the surface morphology.

The microhardness of the cross sections of the treatment sample was measured using a Vickers microhardness tester (Falcon 503) under a 0.1-N load. 7 indentations were performed on each sample, and 5 points average value except both maximum and minimum values were used for hardness.

The surface roughness was measured with the TR-200 profilometer. The change in the weight of samples was determined on the CitizonCY224C electronic analytical balance with an accuracy of ±0.0001 g after washing the samples with distilled water and acetone in an ultrasonic bath to remove traces of salts and subsequently drying.

The wear resistance of samples was studied with the aid of laboratory installation under dry friction conditions with tool alloy steel (wt.%: 0.9–1.2 Cr, 1.2–1.6 W, 0.8–1.1 Mn, 0.9–1.05 C) as counter-body at the normal load of 10 N and sliding speed of 1.555 m/s at the friction path of 1000 m at room temperature. Cylindrical surface of the samples was subjected to wear. Weight loss during the tests was measured with the aid of CitizonCY224C electronic balance after preliminary cleaning of the samples in acetone and subsequently drying. The characteristic morphology of the friction tracks was analyzed using the Micromed MET optical microscope and Quanta 3D 200i scanning electron microscopy with EDX-analysis.

3. Results and discussion

As a result of studying the morphology of the surface and surface layer of steel after CPEB, the presence of structural transformations and the formation of a modified layer 85–250 μm thick (the layer thickness correlates with the duration of treatment) were established. In this case, the surface roughness Ra increases by a factor of 1.5–2.5 in comparison with the untreated sample. The study of the distribution of microhardness in the modified layer showed that the maximum microhardness of 1050 HV and the thickness of the hardened layer of about 500 μm are achieved after boriding for 20 and 30 minutes (figure 1). The tribological behavior of boriding steel correlates with structural changes in the surface layer and the distribution of microhardness (figure 2). The maximum wear resistance of the surface is observed after processing for 30 minutes and is 3.5 times higher than that of untreated steel.

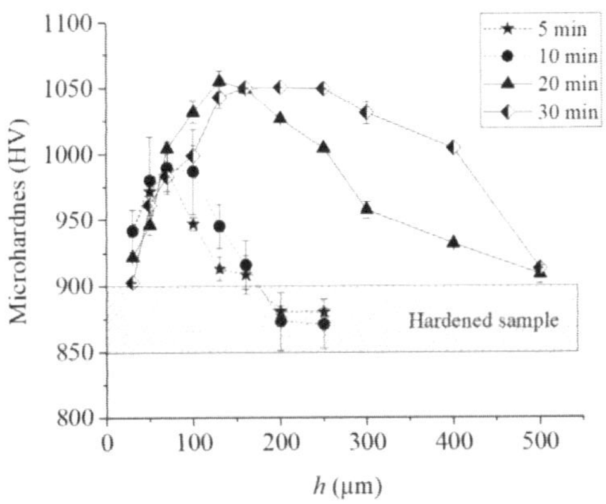

Figure 1. Microhardness distribution in the modified layer after CPEB at 850 °C for different treatment time.

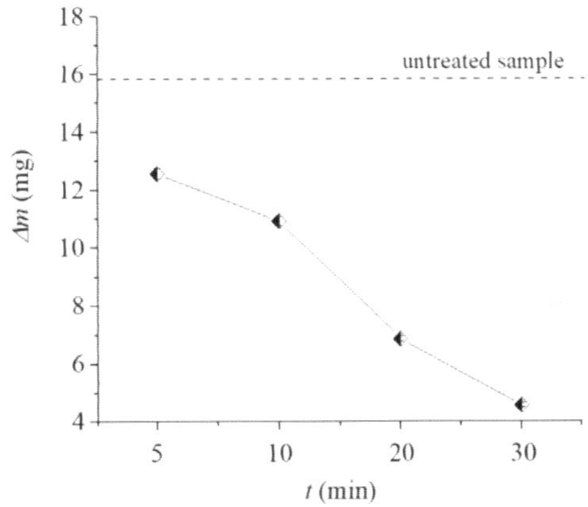

Figure 2. Weight loss of samples during friction after CPEB at 850 °C for different treatment time.

Analysis of the surface of the friction tracks after CPEB indicates the absence of traces of microcutting (abrasive wear), which confirms the good antifriction properties of the modified steel (figure 3). At the same time, EDX analysis indicates an increased content of some chemical elements on the surface (figure 4).

Figure 3. Surface morphology of the wear tracks of sample after CPEB at 850 °C for 30 minute.

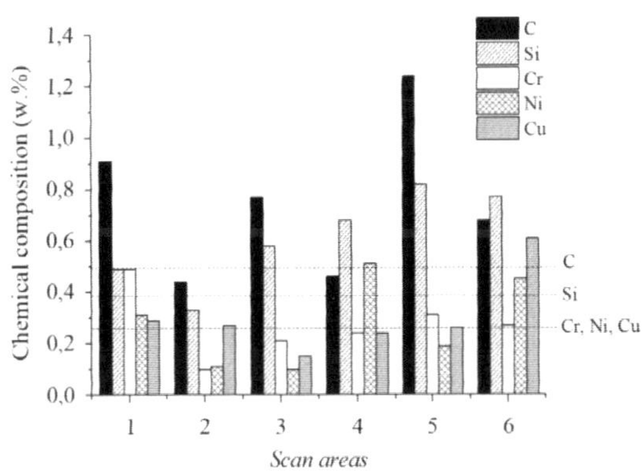

Figure 4. EDX-analysis data in scan areas of figure 3.

APEP of boriding steel (with an optimal CPEB time of 30 minutes) in a solution of ammonium sulfate for 1 minute reduces the average roughness of the sample surface (figure 5), does not significantly reduce the maximum microhardness and thickness of the hardened layer, reduces the coefficient of friction from 0.615 to 0.371 and the weight loss of the samples during friction (figure 6).

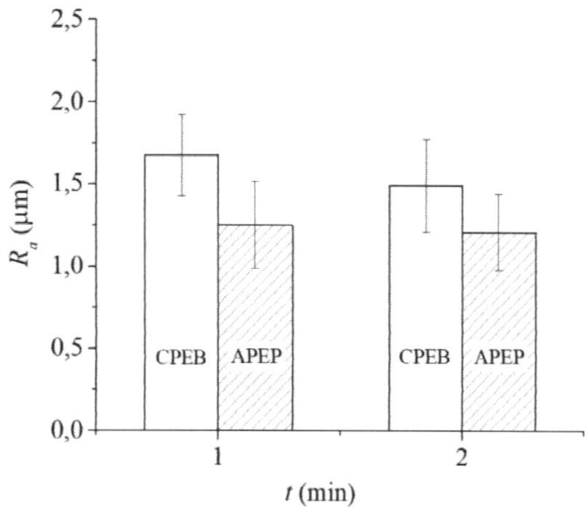

Figure 5. Surface roughness of CPEB samples (850 °C for 30 minute) before and after APEP for 1 and 2 minutes.

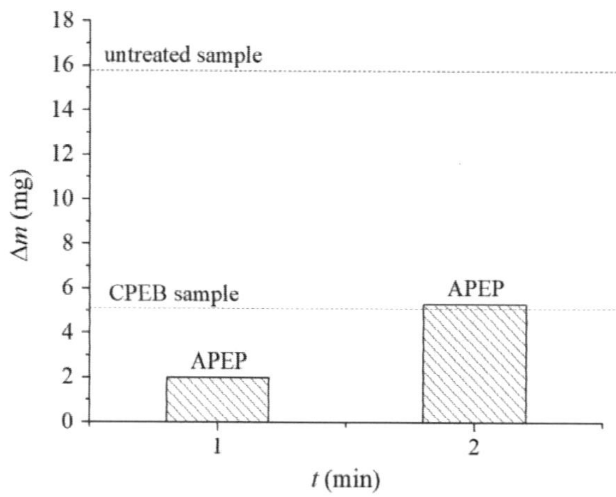

Figure 6. Weight loss of during friction CPEB samples (850 °C for 30 minute) before and after APEP for 1 and 2 minutes.

4. Conclusions

The possibility of anodic polishing of the steel surface after cathodic boriding in electrolysis plasma using the same equipment is shown.

After CPEB, a modified layer up to 250 μm thick with a microhardness of up to 1050 HV is formed with a increase in the surface roughness by 1.5-2.5 times. The wear resistance of the boriding surface is increased by 3.5 times, and the wear mechanism changes from abrasive to polishing.

APEP of the boriding surface leads to a decrease in roughness by 1.4 times and wear by 2.3 times.

Acknowledgments

This work was financially supported by the Russian Science Foundation (Contract No. 18-79-10094) from Kostroma State University.

The authors express their deep gratitude to Prof. S.A. Kusmanov for valuable advice in discussing this work, and also to Dr. T.L. Mukhacheva for help in analyzing tribological studies.

References

[1] Yerokhin A L, Nie X, Leyland A, Matthews A and Dowey S J 1999 *Surf. Coat. Technol.* **122** 73
[2] Aliofkhazraei M, Macdonald D D, Matykina E, Parfenov E V, Egorkin V S, Curran J A, Troughton S C, Sinebryukhov S L, Gnedenkov S V, Lampke T, Simchen F and Nabavi H F 2021 *Appl. Surf. Sci. Advances* **5** 100121
[3] Kusmanov S A, Tambovskiy I V, Sevostyanova V S, Savushkina S V and Belkin P N 2016 *Surf. Coat. Technol.* **291** 334
[4] Bejar M A and Henriquez R 2009 *Mater and Design* **30** 1726
[5] Taheri P, Dehghanian Ch, Aliofkhazraei M, Rouhaghdam A S 2007 *Plasma Process. Polym.* **4** S711
[6] Aliofkhazraei M, Rouhaghdam A S and Gupta P 2011 *Critical Reviews in Solid State and Materials Sciences* **36** 174

A comparative study of the effect of ions of different atomic masses on the magnetoresistive properties of $Co_{90}Fe_{10}$/Cu superlattices

N V Gushchina[1], K V Shalomov[1], V V Ovchinnikov[1,2], N S Bannikova[2], R S Zavornitsyn[2] and M A Milyaev[2]

[1] Institute of Electrophysics, Ural Branch of Russian Academy of Sciences, Amundsena Street 106, Yekaterinburg, 620016, Russia
[2] M.N. Mikheev Institute of Metal Physics, Ural Branch of Russian Academy of Sciences, St. S. Kovalevskoy 18, Yekaterinburg, 620990, Russia

E-mail: guscha@rambler.ru, guschina@iep.uran.ru

Abstract. The effect of inert gas ions with different atomic masses (Ar^+, Xe^+) on the magnetoresistance of $Co_{90}Fe_{10}$/Cu superlattices deposited on a silicon substrate has been investigated by comparison. The Ar^+ ion irradiation has been found to decrease the magnetoresistance more significantly than Xe^+ ion irradiation, which seems to be due to a larger average projective range for Ar^+ (R_p = 5–6 nm) than that for Xe^+ (R_p = 3.3–4.3 nm) and, accordingly, a greater depth of the atom mixing zone (~ (2–3)×R_p) when ions move from the top layers of the superlattice toward the substrate.

1. Introduction

The work is devoted to a comparative study of the effect of irradiation with inert gas ions of different atomic masses (Ar^+ and Xe^+) on the magnetoresistance of currently promising $Co_{90}Fe_{10}$/Cu superlattices. Multilayer metallic magnetic superlattices are high-demand materials for micro and nanoelectronics due to their giant magnetoresistance effect (GMR effect), which is a significant (by tens of percent) decrease in the electrical resistance when a constant magnetic field is applied. The study of the ion irradiation effect on the magnetoresistive properties of superlattices is an actual task, since there is almost no data on the behavior of modern magnetic superlattices under irradiation.

2. Experimental part

Multilayered Co90Fe10/Cu magnetic nanostructures were prepared using a ULVAC MPS-4000-C6 high-vacuum deposition machine (Japan). Pure metals were used as targets-cathodes: Cu – 99.99% (oxygen-free), Ta – 99.9%, as well as Co90Fe10 alloy (deviation from the composition no more than 0.5 at. %) and Ni48Fe12Cr40 alloy (deviation from the composition for chromium no more than 0.5 at. %). The superlattices have the following composition: Si//Ta(10)/(Ni80Fe20)60Cr40 (5)/ /[Co90Fe10(1.4)/Cu(2.3)]12/Ta(3), where Si is the substrate and the layer thickness in nanometers is given in parentheses. Samples 3×8 mm2 in size were prepared from CoFe/Cu superlattice plates to measure their magnetoresistance and irradiate them.

The Co90Fe10/Cu superlattices were irradiated in an ILM-1 ion implanter equipped with a PULSAR-1M ion source based on a low-pressure glow discharge with a hollow cold cathode [1]. Inert

Content from this work may be used under the terms of the Creative Commons Attribution 3.0 licence. Any further distribution of this work must maintain attribution to the author(s) and the title of the work, journal citation and DOI.
Published under licence by IOP Publishing Ltd

gas ions with different atomic masses, Ar^+ (39.95 a.m.u.) and Xe^+ (131.29 a.m.u.), were chosen for irradiation. We used the following ion beam parameters: ion energy $E = 10$ keV, ion current density $j = 100$ μA/cm², and fluences $F_1 = 1.25\cdot10^{15}$ cm^{-2} and $F_2 = 1.0\cdot10^{16}$ cm^{-2}. A ribbon ion beam with a 2×10 cm² cross-section was cut from the ion beam of the circular cross-section using a collimator. The samples were moved under the ion beam at a speed of 2 cm/s to avoid their significant heating upon irradiation. The thin chromel–alumel thermocouple was used for estimating the temperature of the targets in the course of irradiation. It was welded to a witness sample made of thin metal foil with a close coefficient of blackness and the same size as the superlattice samples under study. The thermocouple was read by an Advantech Adam 4000 digital multichannel temperature measurement system. The temperature of the witness sample during irradiation in a continuous regime did not exceed 200°C at the maximum fluence of 10^{16} cm^{-2}.

The magnetoresistance of the superlattices was measured before and after ion irradiation using an AVM-1 automated machine at the "Center for Technology of New Magnetic Materials," Physical-technological infrastructural complex, Institute of Metal Physics, UB RAS. The resistance of the superlattices in the magnetic field (to 19.5 kOe) was measured using a four-contact method using clamped contacts. The magnetic vector and the current in the plane of the film layers were mutually perpendicular to each other during the measurements. The measurements were carried out at room temperature. The magnetoresistance was found as the relative change in the electrical resistance of the material (R), when a magnetic field (H) was switched on: $\Delta R/R(H)=[(R(H)-R_s)/R_s]\cdot100$ %, where R_s is the electrical resistivity of a sample in the case of magnetic saturation, when the magnetic moments of the ferromagnetic layers are parallel; $R(H)$ is the electrical resistivity in magnetic field H; and $(\Delta R/R)_{max}$ is the maximum $\Delta R/R(H)$ value of the field dependence of magnetoresistance.

3. Details of exposure, measurements, and results

Figure 1 shows the magnetoresistance of the $Co_{90}Fe_{10}/Cu$ superlattice before and after Ar^+ и Xe^+ ion irradiation. The Ar^+ ion irradiation decreases the magnetoresistance $(\Delta R/R)_{max}$ of the sample from 22.7 to 17.6 % at fluence $F_1 = 1.25\cdot10^{15}$ cm^{-2} (figure 1a). At fluence $F_2 = 10^{16}$ cm^{-2}, the magnetoresistance of the superlattice is 10.9 %, which is 2.1 times lower than its initial value (figure 1a). The magnetic saturation field also decreases with increasing fluence, and the transition to magnetic saturation becomes more pronounced.

Similarly, the magnetoresistance of $Co_{90}Fe_{10}/Cu$ superlattices also decreases during irradiation with Xe^+ ions (figure 1b). At an ion fluence of $1.25\cdot10^{15}$ cm^{-2}, the magnetoresistance decreases slightly, from 22.7 to 21.8%. The magnetoresistance continues to decrease up to 17.4 % when the fluence increases to $1\cdot10^{16}$ cm^{-2}. Consequently, the magnetoresistance at the same fluences decreases to lower values in the case of Ar^+ ion irradiation than those in the case of Xe^+ ion irradiation.

A decrease in the magnetoresistance of Fe/Cr superlattices under irradiation with 200 and 500 keV gas ions at fluences of ~$5\cdot10^{12} - 5\cdot10^{14}$ cm^{-2} in works [2–5] is associated with the ion-beam mixing of Fe and Cr atoms at interlayer boundaries in the superlattice and a change in the interface state. As shown in [6], 10 keV Ar^+ ion irradiation of Fe/Cr and $Co_{90}Fe_{10}/Cu$ superlattices decreases their magnetoresistance at higher fluences of ~$1.25\cdot10^{15} - 1\cdot10^{16}$ cm^{-2} due to the same processes. However, this case is a partial mixing in depth from the film surface to the substrate, rather than throughout the entire film thickness because of the smaller projective range of Ar^+ ions with energy of 10 keV.

The zone of active mixing during ion irradiation is known to be $(2–3)\times R_p$, where R_p is the average projective range of ions. We can estimate the depth of the mixing zone of atoms from the film surface to the substrate during irradiation with different ions. The SRIM calculation [7] indicates that the average projective range is 5–6 nm for Ar^+ ions and 3.3–4.3 nm for Xe^+ ions at an ion energy of 10 keV in the superlattices under study. Then, the depth of the atom mixing zone in the direction from the film surface to the substrate is 10.0–18.0 nm in the case of Ar^+ ion irradiation and 6.6–13.0 nm in the case of Xe^+ ions. The smaller depth of the zone of atomic mixing during Xe^+ irradiation seems to be responsible for the smaller ion effect on the magnetoresistance of the superlattices under study.

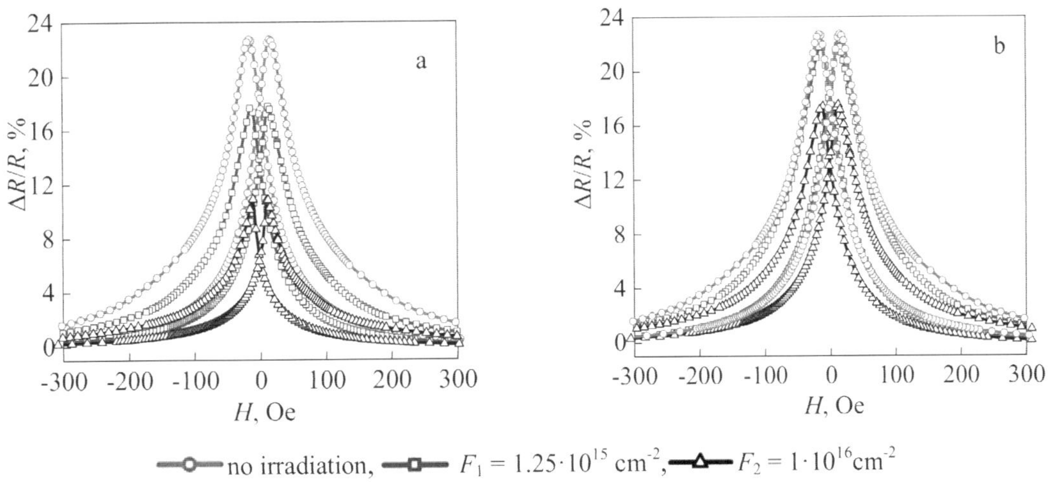

Figure 1. Magnetoresistance curves of Si//Ta(10)/($Ni_{80}Fe_{20}$)$_{60}$$Cr_{40}$ (5)/[$Co_{90}Fe_{10}$(1.4)/Cu(2.3)]$_{12}$/Ta(3) superlattices before and after (a) Ar^+ and (b) Xe^+ ion irradiation at various fluences.

4. Conclusions

The study showed that the 10-keV Ar^+ and Xe^+ ion irradiation (ion current density is 100 μA/cm^2) of the $Co_{90}Fe_{10}$/Cu superlattice on Si substrate at $1.25\cdot10^{15}$ and $1\cdot10^{16}$ cm^{-2} fluences decreased its magnetoresistance with increasing fluence, regardless of the kind of ions. The Ar^+ ion irradiation decreased the magnetoresistance more significantly, which was apparently due to a larger average projective range R_p for Ar^+ than that for Xe^+ and, accordingly, a greater depth of the atom mixing zone in the direction from the film surface to the substrate.

Acknowledgments

This work was supported by the Russian Scientific Foundation, project no. 19-79-20173. The Russian Foundation for Basic Research (project no. 20-42-660018_r-a) supported partially the preparation of $Co_{90}Fe_{10}$/Cu superlattices.

References

[1] Gavrilov N V, Mesyats G A, Nikulin S P, Radkovskii G V, Eklind A and Perry A J 1996 *J. Vac. Sci. Technol.* **A 14** 1050–6
[2] Kopcewicz M, Stobiecki F, Jagielski J, Szyman´ski B, Schmidt M and Kalinowska J 2002 *Hyperfine Interactions* **144/145** 255–9
[3] Kopcewicz M, Stobiecki F, Jagielski J, Szymański B, Schmidt M. Dubowik J and Kalinowska J 2003 *Journal of Applied Physics* **93** 5514–8
[4] Kelly D M, Schuller I K, Korenivski V, Rao K V, Larsen K K, Bottiger J, Gyorgy E M and Dover R B 1994 *Physical Review* **50** 3481–4
[5] Oomi G, Yayoi T, Nakano T, Saito K, Takanashi K, Mitani S, Kohno A and Sagara K 2014 *IEEE transactions on magnetics* **50** 1500304
[6] Gushchina N V, Shalomov K V, Ovchinnikov V V, Bannikova N S, Zavornitsyn R S and Milyaev M A 2020 *Physics of Metals and Metallography* **121** 1271–7
[7] http://www.srim.org/SRIM/SRIM2011.htm

Development of a method for calculating the LTEC anisotropy of Zr alloys

S Ya Betsofen, K A Speransky, A V Bespalov and S S Alexandrova
Moscow Aviation Institute (NRU), Moscow, RF

Email: loz-plasma@yandex.ru

Abstract. The accuracy of determining the LTEC with respect to metals and alloys with an HCP lattice based on inverse pole figures (IPF) is analyzed depending on the number of experimental points on the IPF using three averaging options: (1) taking into account the irregular arrangement of reflections on the stereographic triangle according to Morris; (2) by the multiplicity factor and (3) with the same weight of each orientation. It is shown that when evaluating the LTEC for semi-finished products with a basal texture, 17 reflections on the IPF are sufficient to provide an error of <1% when using Morris averaging and the multiplicity factor; in the case of a prismatic texture, an error of <1% is provided by all three averaging options, while Morris averaging is minimal.

The development of new high-tech materials for the rocket-space and aviation industries, nuclear energy is associated with the creation of technologies that provide the required complex of technological and operational properties of products and structures. Zirconium alloys have found wide application in these industries [1]. A characteristic feature of these alloys is the crystallographic texture, which is formed in the processes of deformation and heat treatment, while for zirconium alloys, the texture almost completely determines the anisotropy of the physical and mechanical properties of semi-finished products and products from them [2–8]. For a quantitative assessment of the type and intensity of the texture of zirconium objects, the Kearns coefficient is widely used [9–11]. One of the most important characteristics of zirconium and its alloys is the LTEC value, which, in contrast to metals with a cubic lattice, has a noticeable dependence on the orientation, since the LTEC value in the direction of the "c" axis of the HCP lattice Zr is equal to α_c = 7.65 10^{-6} deg^{-1}, and the LTEC in the basal plane is $\alpha_a = 5.16 \cdot 10^{-6}$ deg^{-1}.

The most common method for determining the anisotropy of physical properties from texture data is the method of inverse pole figures (IPF) [12–16]. The disadvantage of this method is the limited experimental points on the IPF, while there is no data on the estimation of the error in determining the LTEC and other physical properties, depending on the number of experimental points on the IPF. In [17], the accuracy of estimating the Kearns coefficients and Young's modulus from quantitative IPF was analyzed as a function of the discreteness of the experimental points for various variants of averaging properties on the IPF for HCP alloys, and in this work, such an analysis is implemented for LTEC as applied to Zr and its alloys.

Usually, for HCP alloys, 17 independent reflections are used to construct IPF, which were calculated for titanium, for which there are 17 independent reflections in the X-ray diffraction pattern

Content from this work may be used under the terms of the Creative Commons Attribution 3.0 licence. Any further distribution of this work must maintain attribution to the author(s) and the title of the work, journal citation and DOI.
Published under licence by IOP Publishing Ltd

in copper radiation; for Mg and Zr with long periods, there are already 20 reflections. Therefore, we compared the accuracy of the averaging of properties for 17 and 20 reflections in three different ways.

In the first case, the usual normalization was used, in which all reflections have the same weighting factor (IPF17 and IPF20). The LTEC values (α) were also estimated by summation, taking into account only texture coefficients, but without taking into account the uneven distribution of experimental reflections over the projection sphere.

$$\alpha_n = \frac{1}{n}\sum_{i=1}^{n} P_i\alpha_i, \quad (1)$$

where: n is the number of reflexes (17 or 20).

The LTEC value was calculated using single crystal constants for Zr. Nevertheless, the above calculations can be implemented for any HCP metals and alloys for which there are data on single crystal constants.

$$\alpha_i = \alpha_a + \cos^2\gamma_i(\alpha_c - \alpha_a), \quad (2)$$

where: γ is the angle between the normal to the plane (i = hkl) and "c" axis of the HCP lattice; $\alpha_c = 7.65\cdot10^{-6}$ and $\alpha_a = 5.16\cdot10^{-6}$ deg^{-1} - LTEC constants for Zr

The second method of normalization and averaging took into account the irregular distribution of reflections using the coefficients (IPF17-A and IPF20-A):

$$P_i = \frac{I_i / R_i}{\sum_{i=1}^{n} A_i(I_i / R_i)}\ ; \quad \alpha_n^A = \frac{1}{n}\sum_{i=1}^{n} A_i\alpha_i, \quad (3)$$

where A_{hkl} is the fraction of the surface area of the stereographic triangle around the normal (hkl), bounded by large circles bisecting the angular distances between adjacent normals.

The third method of normalization and averaging took into account the irregular distribution of reflections using the multiplicity factor (N_i), - IPF17-N and IPF20-N:

$$P_i = \sum_{i=1}^{n} N_i \frac{I_i / R_i}{\sum_{i=1}^{n} N_i(I_i / R_i)}\ ; \quad \alpha_n^N = \sum_{i=1}^{n} N_i P_i \alpha_i / \sum_{i=1}^{n} N_i\,. \quad (4)$$

In [18], a methodological approach to assessing the accuracy of determining the Kearns coefficients is proposed. The basis of this approach is the analytical representation of the crystallographic texture, which makes it possible to compare various options for measuring this parameter from real IPFs characterized by a limited number of experimental reflections with the results of simulating similar textures for an arbitrary number of orientations. This approach can also be used to assess the accuracy of determining the characteristics of the LTEC anisotropy. In this case, as in [17, 18], the pole density dependences for model textures can be represented as:

$$P_\gamma^{basal} = 20(1+5\gamma^{1,5})^{-2} \text{ for the basal texture;} \quad (5)$$

$$P_\gamma^{prism.} = 20(1+5(0,5\pi-\gamma^{1,5}))^{-2} \text{ for a prismatic texture.} \quad (6)$$

Figure 1 shows the IPF corresponding to these equations. The LTEC values in accordance with (2) depend only on the angle γ. This makes it possible to carry out averaging over the area of the stereographic triangle by summing "n" trapezoidal sectors, which (figure 2) are limited in height by arcs separated by the angular interval Δγ, and in width - by the sides of the stereographic triangle, the angular distances between which (ρ_i) vary from zero at the pole of the sphere of projections up to 30^0 at the "equator":

$$\Delta S_i = 0,5\Delta\gamma(\rho_i + \rho_{i-1})\,;\ \rho_i = \arccos(\frac{\sqrt{3}}{2}\sin^2\gamma_i + \cos^2\gamma_i);\gamma_i = i\Delta\gamma\ , \qquad (7)$$

where index i ranges from 1 to m, where m ranges from 18 ($\Delta\gamma$ =5^0) to 180 ($\Delta\gamma$ =$0,5^0$).

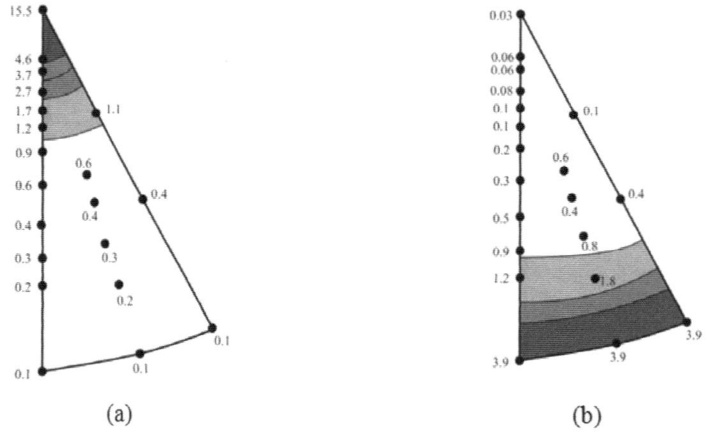

Figure 1. Model IPF20 with basal (a) and prismatic (b) textures, the values of pole densities of which are calculated from equations (3) and (4), respectively.

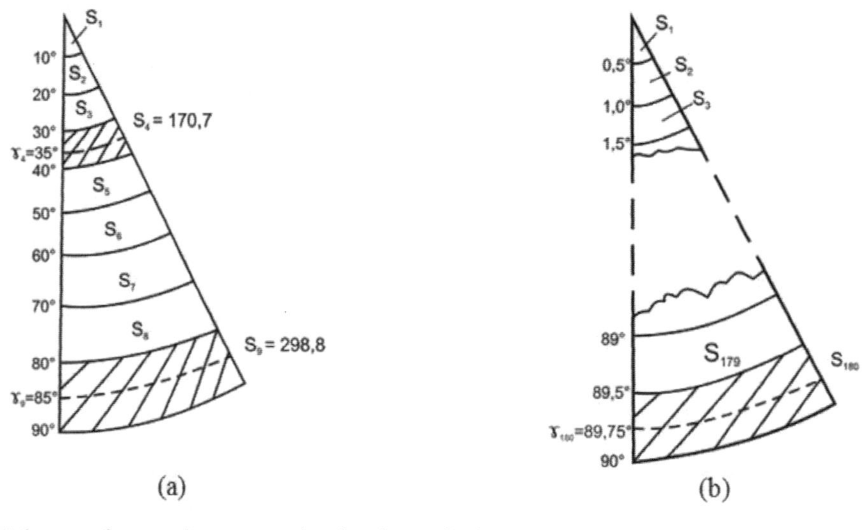

Figure 2. Scheme of averaging properties for theoretical IPF (IPF9, IPF180) at angular distances $\Delta\gamma$ = 100 for IPF9 (a) and 0.50 for IPF180 (b).

This representation of the texture makes it possible to evaluate the accuracy of the IPF method as applied to the estimation of the anisotropy parameters, since on the basis of the analytical function it is possible to absolutely accurately determine the values of the pole densities for the experimentally measured reflections, as well as for any orientations on the pole figure. This approach makes it possible to compare the accuracy of averaging over experimental (IPF17 (20), IPF17 (20) -A, IPF17 (20) -N) and theoretical IPF9, IPF18, IPF45, IPF90, IPF180, since in the latter the discreteness of orientations can be set indefinitely up to such minimum angles between orientations, when the value of the anisotropy parameter will practically not change, and this value can be taken as the true value of the anisotropy parameter for a given texture.

In this case, the error of the IPF method will be equal to the difference between this true value of the anisotropy parameter and the corresponding value obtained for the IPF with 17 reflections, i.e. for such an IPF that we can obtain experimentally. However, for arbitrary experimental textures, the true values of the anisotropy parameters are unknown. In our case, we obtained the values of the pole densities not from the experiment, but calculated directly from Eqs. (3), (4) precisely for the texture for which the true value of the anisotropy parameter is known. This provides a unique opportunity to evaluate the error in determining the anisotropy parameters for any type of texture and for various averaging techniques.

Thus, the calculation procedure for estimating the LTEC includes determining the pole density for an arbitrary IPF according to equations (3), (4), calculating the LTEC values for the corresponding points on the IPF, determined by the polar angle γ_i, and averaging these values taking into account the pole density at a given IPF point :

$$f_{\Delta\gamma} = m\sum_{i=1}^{m}\Delta S_{\gamma_i} P_{\gamma_i}\cos^2\gamma_i / \left(\sum_{i=1}^{m}\Delta S_{\gamma_i}\sum_{i=1}^{m} P_i\right), \qquad (8)$$

where m is the number of sectors in the IPF.

In figure 2 shows averaging schemes for the case m = 9 and $\Delta\gamma = 10^0$ and m = 180 for $\Delta\gamma = 0.5^0$. The relative error was estimated by the ratio:

$$100(\alpha m - \alpha 180)/\, \alpha 180, \qquad (9)$$

where: α_m- LTEC for m = 18, 45, 90; α_{180}- LTEC for m = 180

In [18], it was shown that the averaging of the IPF17-A gives significantly more accurate values of the Kearns coefficients as compared to the averaging of the IPF17, which does not take into account the irregularities of reflections on the projection sphere. Figure 3 shows a comparison of the errors in determining the LTEC depending on the averaging method and the number of experimental points on the IPF, respectively, for alloys with basal and prismatic textures.

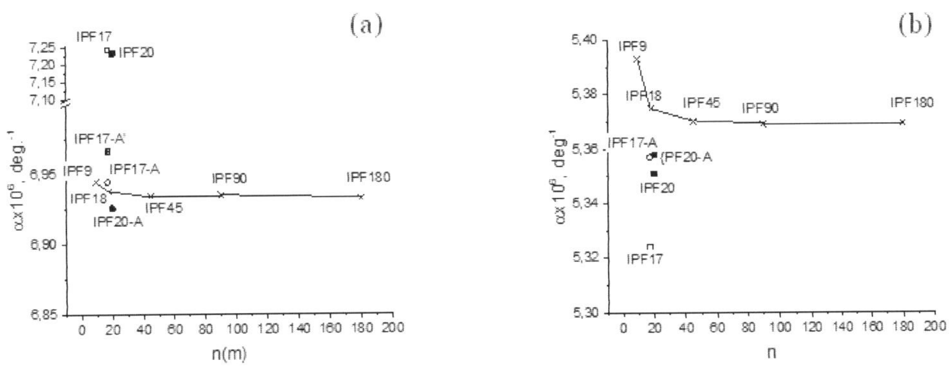

Figure 3. Comparison of the accuracy of calculating the LTEC depending on the type of averaging and the number of experimental points on the IPF for the basal (a) and prismatic (b) textures.

1. The estimation of the accuracy of determining the LTEC was carried out using three variants of averaging on the IPF, taking into account the irregular arrangement of reflections on the stereographic triangle according to Morris, according to the multiplicity factor and with the same weight of each orientation.

2. It is shown that when evaluating the LTEC for semi-finished products with a basal texture, 17 reflections on the IPF are sufficient to provide an error of <1% when using Morris averaging and the

multiplicity factor; in the case of a prismatic texture, an error of <1% is provided by all three averaging options, while the Morris averaging error is minimal.

Acknowledgments
The authors are deeply grateful to A.A. Lozovan for a discussion of the results and valuable comments on the work.

References

[1] Breeders S Yu, Zuev L B, Kotrekhov V A 2012 *Metallurgical issues of the production of products from zirconium alloys* Novosibirsk: Nauka 252 p
[2] Murty K L and Charit I 2006 *Progress in Nuclear Energy* **48** 325–59
[3] Yanzhang L, Kim H G, Park J Y et al. 2013 *Rare Metal Mater. Eng.* **42** 4 667–72
[4] Gallaugher M, Peykov D, Brodusch N et al. 2013 *J. Nuclear Materials* **442** 116–23
[5] Mosbrucker P, Daymond M R and Holt R A 2010 *In situ studies of variant selection during the* $\alpha\rightarrow\beta\rightarrow\alpha$ *phase transformation in Zr-2,5Nb* (Zirconium in the Nuclear Industry: Proc. 16th Int. Symp. ASTM) **8** 1 pp 195–215
[6] Fong R W L 2013 *J. Nuclear Materials* **440** 288–97
[7] Azhanzha V M, Kovtun G P, Malykhin D G et al. 2002 *Problems of Atomic Science and Technology* **77** 2 86–93
[8] Perlovich Yu A, Grekhov M M, Isaenkova M G et al. 2004 *Problems of Atomic Science and Technology* **85** 3 59–65
[9] Kearns J J 2001 *J. Nuclear Materials* **299** 171–4
[10] Perlovich Yu A and Isaenkova M G 1999 *On the determination of the Kearns parameters used to characterize the crystallographic texture of pipes made of Zr-based alloys* (Proceedings of the conference "Problems of Zirconium and Hafnium in Nuclear Power Engineering". Kharkov) pp 89–90
[11] Mani Krishna K V, Srivastava D, Dey G K et al. 2011 *J. Nuclear Materials* **414** 492–7
[12] Morris P R 1959 *Journal Applied Physics* **30** 595–8
[13] Gogishvili O Sh, Kononov G G, Lavrinenko I P and Lalykin S P 1981 *Zavodskay laboratoria* **7** 32–4
[14] Kosenkov V M 2006 *X-ray diffraction in reactor materials science* 2nd ed. UlSU, Ulyanovsk 168 p
[15] Borodkina M M and Spektor E N 1981 *X-ray analysis of the texture of metals and alloys* Moscow: Metallurgy 272 p
[16] Betsofen S Ya, Grushin I A, Gordeeva M I, Speransky K A 2021 *Deformation and destruction of materials* **8** 21–9
[17] Dzunovich D A, Betsofen S Ya and Panin P V 2016 *Deformation and destruction of materials* **11** 8–16

Plasma and photon technologies against bio-factors, dangerous to human being

O D Volpian[1], R V Dronsky[2], Yu A Obod[2], A I Kuzmichev[3], O V Andrienko[3] and K V Alybin[4]

[1] LLC "Scientific and Production Complex Fotron-Auto", Moscow 115191, Russia
[2] Scientific and technological centre of unique instrument-making of RAS, Moscow 117342, Russia
[3] Igor Sikorsky Kyiv Polytechnic Institute, Kiev 03056, Ukraine
[4] Scientific and Production Corporation "Precision Instrumentation Systems", Moscow 111024, Russia

E-mails: fotron-auto@mail.ru, kuzmichev-kpi@ukr.net

Abstract. Possible plasma and photonic methods and devices for monitoring and preventing dangerous infections and human diseases are presented. In experiments with different types of atmospheric pressure discharges in different gases, the significant bactericidal effect was found. The prototype of device based on the method of absorption spectroscopy for detecting human diseases by biomarkers in the exhaled air has been proposed and tested. The importance of the plasma technology of deposition of coatings by magnetron sputtering for the creation of anti-covid masks and high-quality optics (mirrors) for photon monitoring devices is emphasized.

1. Introduction

Since biological hazards (as bio-factors) constantly lie in wait for humans, an example of which is the global pandemic COVID-19, scientists and engineers are constantly looking for ways to combat biohazards, proposing new approaches and creating new devices. Plasma and photonic (optical) technologies have great potential along this path [1–3]. Plasma methods are attractive from several sides: firstly, they make it possible to generate highly effective germicidal agents with fairly simple and relatively cheap devices, and secondly, plasma production technologies make it possible to manufacture effective devices for combating bio-hazards by other methods, in particular, using photonics. Photonic and optical technologies are at the heart of prevention, monitoring and detection of diseases, as well as disinfection and sterilization. Photon measurements are used in devices for remote molecular analysis and thermometry of people. Examples of such devices are multispectral cameras with high quantum efficiency, laser diodes and LEDs in visible and ultraviolet light, infrared bolometric arrays, narrowband optical filters, and broadband multispectral optical spectrometers. This work is devoted to the consideration of the aspects of plasma and photonic technologies that we are developing.

Content from this work may be used under the terms of the Creative Commons Attribution 3.0 licence. Any further distribution of this work must maintain attribution to the author(s) and the title of the work, journal citation and DOI.
Published under licence by IOP Publishing Ltd

2. Investigation of the bactericidal effect of atmospheric pressure discharges

In recent decades, much attention has been paid to the study of the effect of gas discharges and plasma generated in them on biological objects [1, 2]. This effect is due to the physical and chemical properties of plasma, which are determined by the fact that the plasma contains a variety of active particles (reactive oxygen and nitrogen species (RONS), hydrogen peroxide additives, free radicals, excited atoms and molecules, electrons, various ions, even active species of inert gases and radiation of photons from UV to IR wavelength range) in a wide energy range (0.05-0.1 eV for atomic particles, for ions and electrons - up to 10 keV). Moreover, to influence biological objects, as a rule, non-equilibrium non-thermal plasma is used, in which the role of high-energy electrons is important. Reaction plasma is used in biomedical equipment for the treatment of various diseases, preparation of liquid and gaseous medicinal media, disinfection and sterilization, as well as for purification, deodorization and disinfection of air. This direction in plasma technology is called as "Plasma in Medicine". The proof of its relevance is the creation of the new journal "IEEE Transactions on Radiation and Plasma Medicine" [1]. The ozone treatment of various biological objects, including agricultural products, as well as drinking water and waste water, is close to this direction. In these technologies, plasma discharges are used to generate ozone.

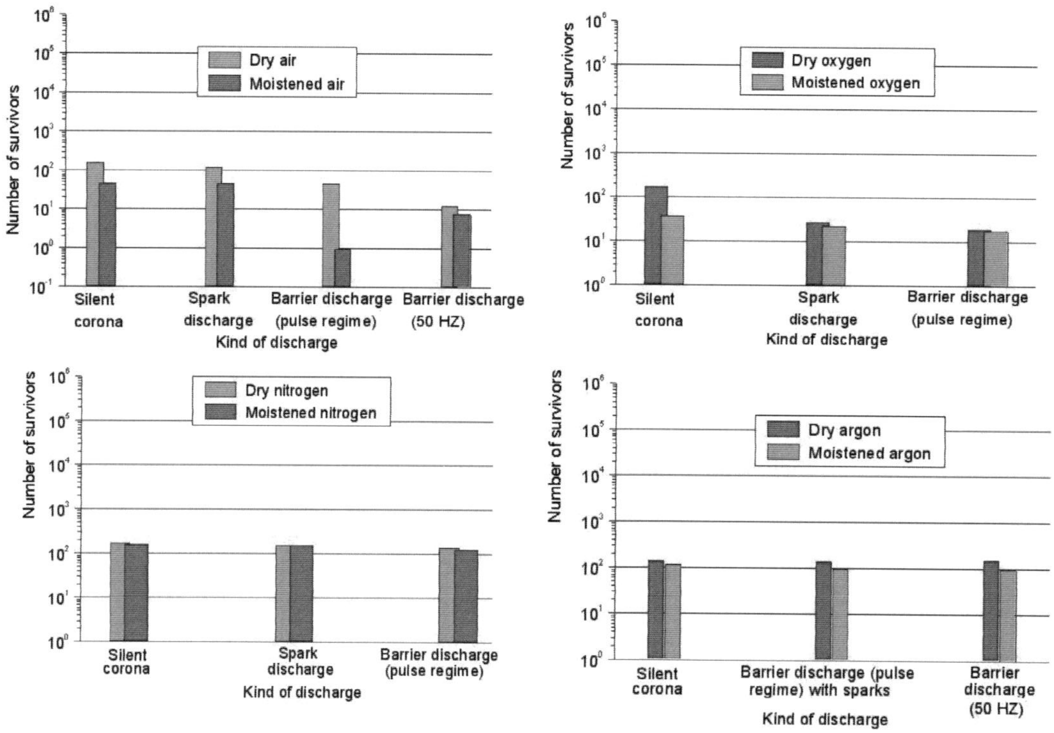

Figure 1. Survival of bacteria spores Bac. Stearothermophilus in 30 min. The atmospheric pressure discharges are sustained in different gaseous media, which were created in a glass chamber with a diameter and height of 20 cm. Discharge power related to the volume of the chamber (W/cm3): for corona – 1·10–4, for spark discharge – 6,3·10–3, dielectric barrier discharge (pulse mode) – 4·10–3, barrier discharge (power supply is by alternating voltage 50 Hz) – 1.5·10–3. The initial number of spores is 8.6·105.

Our studies of atmospheric pressure discharges have shown that the chemical and germicidal effects of exposure to biological objects are inherent in all types of the discharges. See figure 1 with data for such discharges as silent (streamerless) corona, spark discharge and dielectric barrier ones. After exposition of *Bac. Stearothermophilus* spores in the discharge chamber, the number of the survivors is very small. Moistening gases diminishes the number of the survivors due to acids generation. The maximum effect is reached with barrier discharge in air.

The coronavirus (COVID-19) pandemic has sparked a surge of interest in the research and development of plasma devices to target this virus [1]. The positive results of studies with other viruses are encouraging. The exact mechanism of the virucidal effect is still being investigated, but there is the reason to assume that the activity of viruses is suppressed as a result of the oxidation by the active species of its outer spines. Currently, the most realistic plasma technology is the fight against the spread of viruses in the atmosphere of confined spaces with people by equipping ventilation systems with plasma devices while taking all measures of safe operation.

3. Plasma for the production of materials and devices for fighting infections

The second aspect of the application of plasma technologies against coronavirus is the use of plasma methods, in particular, magnetron sputtering [4, 5] for applying active antiviral materials to the surface of masks to prolong their sterile action. It is known that coronaviruses quickly die on the surface of copper alloys, in particular, of the composition Cu-Zn. There is also data on the antiviral activity of Ag films and nanoparticles and TiO_2 photolytic films, which are effectively applied to roll materials by the high-productivity magnetron method. The same method with the reactive plasma assistance [4, 5] is promising for the manufacture of high-quality narrow-band multilayer filters for fluorescent systems for molecular diagnostics of coronavirus infection using a technique called as real-time reverse transcription polymerase chain reaction [3]. It is based on very sensitive spectroscopic detection of extremely small amounts of viral material in a swab from a patient's nose or throat.

4. Photonic device for monitoring human diseases by gas biomarkers

For effective treatment of patients, early detection of signs of disease is desirable, for example, by biomarkers (BM) in the exhaled air [6, 7]. We are developing an optical device prototype that allows determining the presence of low concentrations of BM molecules in the exhaled air, which are derivatives of pathologies in the human organism and indicate diseases in initial stages in the absence of other signs. It is assumed that the reliability of the diagnosis of the disease in the early stages by BM must be sufficiently high. The operation of the device may be based on the method of gas analytical spectroscopy by light absorption in a high-Q optical resonator after wave excitation by external light (may be pulsed one) [7]. Figures 2a depicts the variant of scheme of such device. The device contains a gas cuvette through which the air with BM additive is passed. If BM is detected one makes the conclusion about the decease presence. In the given work the gas CO was taken as a model of BM. Clinical studies have shown, as it is known from literature, for example in [7], the increase in the concentration of exhaled CO is observed in the next diseases: the infection with *Helicobacter pylori* bacteria, liver dysfunction, including cirrhosis, bacterial overgrowth, pancreatic dysfunction, malabsorption, bile metabolism, glucose metabolism, anemia, hemoglobinuria, respiratory tract infection, asthma, etc.

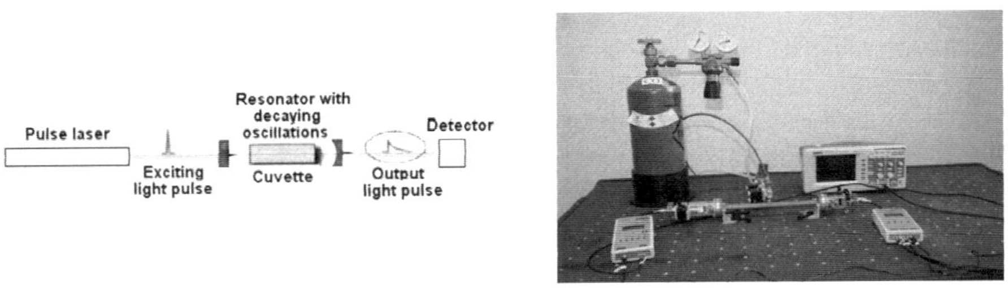

(a) (b)

Figure 2. Scheme of gas spectroscopy (a) based on the light absorption and decay of a light wave in the high-Q resonator and the photo of experimental spectroscopic set-up (b).

The experimental device prototype has been built in the accordance with the scheme of figure 2a and its photo is shown in figure 2b. The resonator has been adjusted for the chosen type of BM, namely CO, and for registration of the first overtone of vibration spectra at 2.33 μm. Testing the device showed the minimal detectable CO content in Ar medium was about 1 ppm with error of order of 20 %. The further experiments with the device are continued.

To carry out the diagnosis of COVID-19, the search for a complex of BM is carried out for providing the diagnostic reliability of over 90 %, as well as commercial device designs for stationary and mobile applications are to be developed. Preclinical and clinical trials must be fulfilled, too.

Note, the device must use high quality, very low loss, narrow band mirrors to ensure high Q of the optical cavity and sensitivity of gas analysis. For the manufacture of such mirrors with a precision multilayer nanostructure, the most suitable coating method is on the base of magnetron sputtering with reactive plasma assistance [4]. The technological facilities for such coating method are described in our paper [5].

5. Conclusions

The results obtained in this work and the literature data confirm the great potential of plasma and photonic technologies in the fight against infections and diseases that are dangerous to humans. In particular, various types of atmospheric pressure discharges in atmospheric air and other gases can be recommended for disinfecting air in ventilation systems, but subject to measures to protect a person from exposure to harmful products of plasma processing. Plasma methods in the magnetron sputtering form and with reactive plasma assisting are recommended for applying various coatings with a germicidal effect on objects used by humans, and for producing high-quality optics for devices that detect early signs of human diseases using gas biomarkers. Experiments have confirmed the possibility of creating the quick non-contact method for diagnosing human diseases based on the presence of biomarkers in the exhaled air, but it is necessary to search for suitable biomarkers for diagnosing diseases such as COVID-19.

Acknowledgements

The work was carried out within the framework of the project "Development of large-scale scientific and scientific-technological projects in priority research areas of the national project «Science and Universities of Russia»", funded by the Ministry of Education and Science of the Russian Federation.

Authors thank V. Tsiolko and V. Bazhenov for their valuable participation in plasma experiments.

References

[1] Bekeschus S, Kramer A, Suffredini E, von Woedtke T and Colombo V 2020 *IEEE Trans. Rad. Plasma Med. Sci.* **4** 391–9

[2] Wolf R A 2013 *Atmospheric pressure plasma for surface modification* (Beverly, MA: Scrivener Publishing)

[3] Baer T M and Baer C E 2020 *Optics & Photonics News* **31** 18–21

[4] MacLeod H A 2010 *Thin-film optical filters* (Boca Raton, FL: CRC Press, Taylor & Francis Group)

[5] Volpian O D, Kuzmichev A I and Churikov D V 2019 *Journal of Physics: Conference Series* **1281** 012090 (1–4)

[6] Waltz E 2021 *Spectrum* **58** (10) 10–1

[7] Liang Q, Chan Ya-C, Changala P B, Nesbitt D J, Ye J and Toscano J 2021 *Proc. Nat. Acad. Sci.* **118** (40) e2105063118 (1–6)

Evaluation of tribotechnical properties of modified surfaces after plasma electrolytic treatment

T L Mukhacheva, T M Kalinina and S A Kusmanov

Department of Physical, Mathematical and Natural Science, Kostroma State University, 17 Dzerzhinskogo Street, Kostroma 156005, Russia

E-mail: sakusmanov@yandex.ru

Abstract. The article considers the study of the effect of plasma electrolytic processing on the tribotechnical characteristics of medium-carbon steel. Friction tests were carried out in dry friction mode. Electron microscope and profilometer were used to study the friction tracks. For a comprehensive assessment of the quality of the modified surface layer, the Kragelsky-Kombalov complex parameter was calculated. It was found that plasma electrolytic treatment leads to a decrease in the coefficient of friction and weight wear in comparison with hardened and untreated steel. It has been determined that the mechanism of wear of samples after plasma electrolytic treatment is fatigue wear at boundary friction and plastic contact.

1. Introduction

Chemical heat treatment of steel products occupies an important place among the methods of increasing their wear resistance. Many publications have shown the possibility of effectively increasing the wear resistance of steel parts by means of their plasma electrolytic saturation with nitrogen, carbon or boron. Cathodic plasma electrolytic nitriding reduces the wear rate of G3500 cast iron by 2.5 times, and S0050A steel by 3 times under dry friction conditions [1]. However, the effect of electrical discharges during cathodic treatment led to an increase in the coefficient of friction. Similar results were obtained for plasma electrolytic nitriding high-speed steel R6M5 [2] or alloy steel 34CrNi1Mo [3], where the resistance to abrasive wear is reduced by 1.5 times. Positive results were obtained with cathodic plasma electrolyte boriding of steel H13 [4] Anodic processes are characterized by a decrease in surface roughness due to the anodic dissolution of the sample, which makes it possible to reduce the friction coefficient and the intensity of wear [5].

The disadvantage of the existing works is the lack of study of the wear mechanism and a comprehensive study of the effect of surface microgeometry on the tribotechnical characteristics of treated materials. This paper proposes to consider this issue.

2. Materials and methods

Cylindrical samples of medium-carbon steel (0.45 wt.% C) 10 mm in diameter and 10 mm in height were subjected to various types of plasma electrolytic processing.

Anodic plasma electrolytic nitrocarburising was carried out in an aqueous solution of ammonium chloride (10 wt.%) and carbamide (20 wt.%) at the sample temperature of 750 °C for 7 min, followed by quenching in the same electrolyte by turning the electrolyzer voltage off.

Content from this work may be used under the terms of the Creative Commons Attribution 3.0 licence. Any further distribution of this work must maintain attribution to the author(s) and the title of the work, journal citation and DOI.
Published under licence by IOP Publishing Ltd

Anodic plasma electrolytic nitriding was carried out in an aqueous solution of ammonium nitrate (5 wt.%) and ammonium chloride (15 wt.%) at the sample temperature of 850 °C for 5 min, followed by plasma electrolytic polishing in the aqueous solution of ammonium sulfate (5 wt.%) under voltage of 300 V and the solution temperature of 80 °C for 1 min.

Cathodic plasma electrolytic boriding was carried out in an aqueous solution of boric acid (3 wt.%) and ammonium chloride (10 wt.%) at the sample temperature of 850 °C for 30 min, followed by plasma electrolytic polishing in the aqueous solution of ammonium sulfate (5 wt.%) under voltage of 325 V and the solution temperature of 70 °C.

The friction scheme "shaft-block" was used in friction tests (figure 1). The counter body was made of tool alloy steel (wt.%: 0.9–1.2 Cr, 1.2–1.6 W, 0.8–1.1 Mn, 0.9–1.05 C) in the form of a plate with a semicircular notch 10 mm in diameter, enclosing the surface of the sample. The sample was mounted on a shaft driven by an electric motor. The counter body was mounted on a platform sliding along cylindrical guides. The platform was moved using a pneumatic cylinder. The cylinder, guides and platform were able to rotate with the pendulum. The pendulum shaft was located coaxially with the sample. Such a scheme makes it possible to preserve the common rotation axis for the sample and the counter body as they wear out and to avoid the influence of misalignment on the results of measurements of the frictional moment.

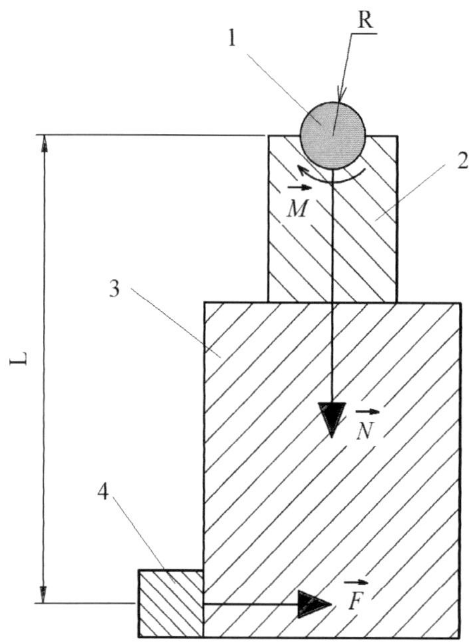

Figure 1. Friction scheme. 1 – sample; 2 – counter body; 3 – pendulum; 4 – strain gauge. M – frictional moment, N – force acting on the counter body and the pendulum from the side of the sample, F - force acting on the pendulum from the strain gauge, L – distance from the axis of rotation to the axis of symmetry of the strain gauge.

Friction tests were carried out in dry friction mode under a load of 10 N. The sliding speed of the sample along the counter body was 1.555 m/s. The friction path was 1000 m.

The friction tracks images were obtained using a Quanta 3D 200i scanning electron microscope. The characteristics of the microgeometry of the friction track surface were investigated on a TR-200 profilometer.

For a comprehensive evaluation of the quality of the modified surface layer, the Kragelsky-Kombalov complex parameter was used, since, in addition to the height parameter, it includes the parameters of the distribution of irregularities along the height and the average radius of curvature of microprotrusions.

$$\Delta = \frac{R_{\max}}{rb^{\frac{1}{\nu}}}, \quad (1)$$

where $R_{\max}$ is the maximum height of irregularities; r is the radius of microroughness, determined by modelling the protrusions with bodies of double curvature:

$$r = \frac{9R_a^2 S_m^2}{128\left(5,5R_a - R_p\right)^3} \tag{2}$$

The constants v and b are determined oexperimentally from the measurement results of the profilograms of a rough body [6]:

$$v = 2l_m\left(\frac{R_p}{R_a}\right) - 1 \tag{3}$$

$$b = l_m\left(\frac{R_{\max}}{R_p}\right)^v, \tag{4}$$

where R_a is the arithmetic mean deviation of the profile; R_p is smoothing height or distance from the line of protrusions to the centre line within the base length [5], l_m is relative reference length of the profile at the level of the centre line.

3. Results and discussion

Parameter (1) determines the load-carrying capacity of the roughness profile. The results of its calculation are presented in table 1.

Table 1. Kragelsky-Kombalov criterion (Δ), wear resistance and frictional properties.

Processing type	Δ	Average friction coefficient	Maximum wear (mg)
Anodic plasma electrolytic nitrocarburising	0.253±0.004	0.445±0.007	3.02±0.03
Anodic plasma electrolytic nitriding with plasma electrolytic polishing	0.158±0.002	0.526±0.008	1.19±0.02
Catodic plasma electrolytic borizing with plasma electrolytic polishing	0.161±0.002	0.388±0.006	5.31±0.11
Furnace quenching	0.398±0.006	0.622±0.009	8.02±0.15
No processing	0.892±0.013	0.798±0.014	21.90±1.11

Plasma electrolytic processing with subsequent plasma electrolytic polishing allows obtaining the maximum bearing capacity of roughness at the friction contact, which is evident from the minimum values of the Δ criterion. The bearing capacity of the profile is the smallest for untreated and quenched samples.

The relief observed in the photographs of the friction tracks (figure 2, 3 and 4) corresponds to the calculated values of the Δ criterion in table 1. The distribution of the material along the height of the rough layer and the distribution of the tops of microprotrusions and their height along the base length are more even and smooth on the friction track of the sample after plasma electrolytic processing (figure 2). In the bright areas, plastically deformed microvolumes of the metal are formed, which, with repeated exposure, lead to the displacement of the material with permanent deformation after the passage of the microprotrusion.

The properties of the frictional contact, specified, in particular, by the Kragelsky-Kombalov criterion, largely determine the processes of friction and wear. Samples after all types of plasma electrolytic processing are characterized by a lower coefficient of friction and weight wear compared to quenching and untreated steel (table 1).

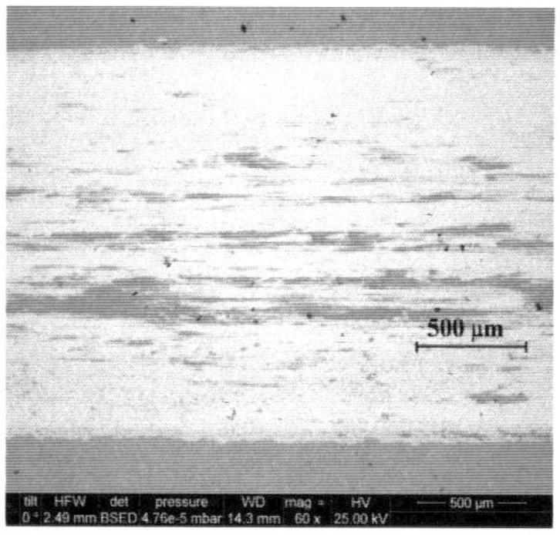

Figure 2. SEM image of the friction track of a nitrocarburised sample.

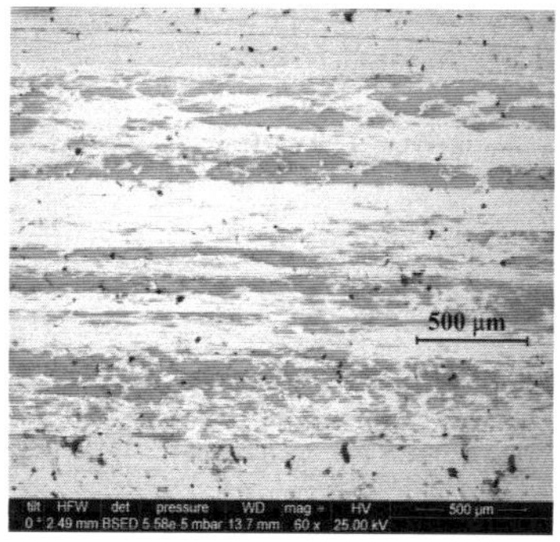

Figure 3. SEM image of the friction track of a furnace-quenched sample.

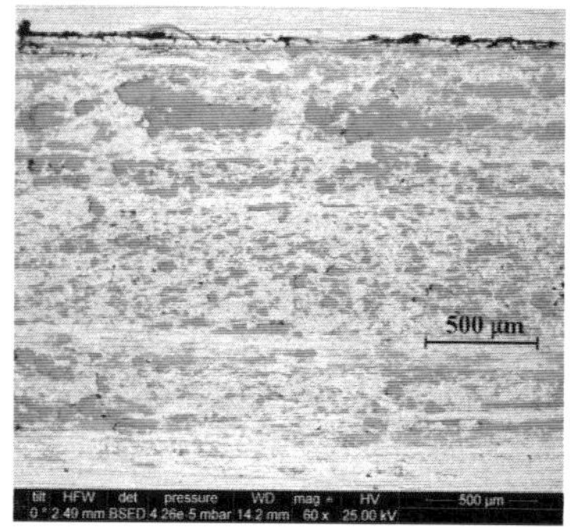

Figure 4. SEM image of the friction track of an untreated sample.

The type of wear of the samples modified with plasma electrolytic processing is fatigue wear at boundary friction and plastic contact. With classical quenching or no processing, the plastic contact is replaced by microcutting.

4. Conclusions

It has been established that plasma electrolytic treatment with subsequent plasma electrolytic polishing allows obtaining the maximum bearing capacity of the roughness at the friction contact, as evidenced by the minimum values of the Kragelsky-Kombalov criterion. The bearing capacity of the profile is the smallest for untreated and quenched samples.

A decrease in the coefficient of friction and mass wear after all types of plasma electrolytic treatment in comparison with hardened and untreated steel has been established.

It is shown that the type of wear of samples modified by plasma electrolytic treatment is fatigue wear at boundary friction and plastic contact. With or without conventional hardening, the plastic contact is replaced by microcutting.

Acknowledgments

This work was financially supported by the Russian Science Foundation (Contract No. 18-79-10094) from Kostroma State University.

References

[1] Nie X, Wang L, Yao Z C, Zhang L and Cheng F 2005 *Surf. Coat. Technol.* **200** 1745
[2] Skakov M, Rakhadilov B, Scheffner M, Karipbaeva G and Rakhadilov M 2013 *Appl. Mech. Mater.* **379** 161
[3] Skakov M, Verigina L and Scheffner M 2015 *Appl. Mech. Mater.* **698** 439
[4] Taheri P, Dehghanian Ch, Aliofkhazraei M and Sabour Rouhaghdam A 2007 *Plasma Process. Polym.* **4** S711
[5] Mukhacheva T L, Belkin P N, Dyakov I G and Kusmanov S A 2020 *Wear* **462-463** 203516
[6] National standard of the Russian Federation. Geometrical Product Specifications (GPS). Moscow, Standartinform, 2015 http://docs.cntd.ru/document/1200116337

Temperature distribution in a cylindrical lengthy workpiece under plasma electrolytic heating

S Yu Shadrin, A V Zhirov and T L Mukhacheva

Department of Physical, Mathematical and Natural Science, Kostroma State University, 17 Dzerzhinskogo Street, Kostroma 156005, Russia

E-mail: syushadrin@yandex.ru

Abstract. The article focuses on developing a model of a stationary temperature field inside a semi-infinite cylindrical sample partly immersed in electrolyte. Temperature calculation is carried out by solving a heat transfer equation separately for the immersed and protruding parts of the sample. The heat flux is set on the outer area boundaries; meanwhile heat flux density for the immersed part is linearly dependent on the vertical coordinate. Within the framework of the model, vertical and radial temperature gradients are worked out both in protruding and immersed parts of the anode. It has been established that the vertical coordinate of the sign reversal point of the heat flux density in the immersed part depends on heat exchange conditions in the protruding part.

1. Introduction

Plasma electrolytic heating is one of rapid methods of thermochemical treatment applicable for hardening, diffusive saturation of the surface layer with light elements, surface cleansing or polishing, coating deposition etc. One of the most important parameters of the process of metal heating with electrolyte plasma is the temperature of an anode. By now, there is empirical evidence of an influence of various factors on this value, furthermore theoretical calculation models for short anode temperature under its various heating conditions have been designed [1, 2]. In the case of a fairly lengthy treatment area, more accurate knowledge of stationary temperature distribution in the work piece volume is crucial, which would allow to predict the influence of heating no uniformity on the diffusion processes in the surface layer.

One of the first models of temperature distribution in the submerged part of an anode was suggested in [3], where stationary temperature distribution was calculated in a heated semi-infinite cylinder under the following simplifying assumptions. The sample was divided into two areas: immersed and protruding. For each area, heat transfer equation was solved in the cylindrical coordinate system. It was assumed that the temperature at the border of the immersed part and the vapour gas envelope was linearly changing, whereas the temperature of the cylinder side showing over the solution surface was changing exponentially, analogous to temperature distribution under convective heat exchange in a slender column. The heat flux on the ends of the sample was neglected; moreover temperature continuity and heat flux conditions at the boundaries of immersed and protruding parts were added. The assigned problem has an accurate solution obtained via integral transformation method. The solution shows weak radial dependence on the temperature on the end of the immersed part and in the protruding part distant from the electrolyte boundary.

Content from this work may be used under the terms of the Creative Commons Attribution 3.0 licence. Any further distribution of this work must maintain attribution to the author(s) and the title of the work, journal citation and DOI.
Published under licence by IOP Publishing Ltd

In further studies [4], both direct and indirect experimental data showed a different temperature distribution in a submerged part of a heated sample, associated with different heat flux density distribution at the boundary of an anode and a vapour gas envelope. Thus the aim of this work is modernization of calculation methods for temperature distribution in a lengthy cylindrical anode by means of assigning different boundary conditions.

2. Calculation method

A semi-infinite cylindrical anode sample is divided into two parts: immersed and protruding. For each part, a heat transfer equation is solved in the cylindrical coordinate system with its own boundary conditions (figure 1).

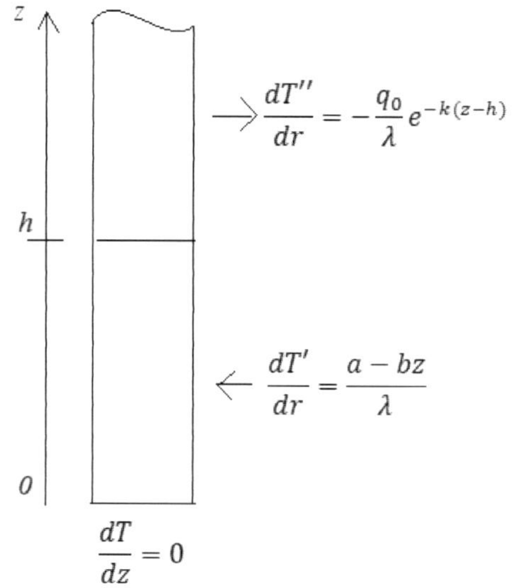

Figure 1. Scheme of temperature calculating.

As in [3], the heat flux on the end of the sample is neglected and the boundedness condition is used for solving at infinity. Heat flux densities on the sides of the sample are set; there are linear relations between the heat flux density and the vertical coordinate on the immersed part. For the protruding part, the known temperature distribution at convective heat exchange in a semi-infinite rod is applied. Moreover, area matching conditions are added at $z = h$. Then the following mathematical formulation of the boundary value problem is obtained:

Area I Area II

$$\frac{\partial^2 T_1}{\partial r^2} + \frac{1}{r}\frac{\partial T_1}{\partial r} = 0 \qquad \frac{\partial^2 T_2}{\partial r^2} + \frac{1}{r}\frac{\partial T_2}{\partial r} = 0$$

$$z = 0 \quad \frac{\partial T_1}{\partial z} = 0 \qquad z \to \infty \quad \frac{\partial T_2}{\partial z} = 0 \tag{1}$$

$$r = R \quad \frac{\partial T_1}{\partial r} = \frac{a - bz}{\lambda} \qquad r = R \quad \frac{\partial T_2}{\partial r} = -\frac{q_0}{\lambda}\exp\left(-k\left(z-h\right)\right)$$

$$z = h \quad T_1 = T_2 \quad \frac{\partial T_1}{\partial z} = \frac{\partial T_2}{\partial z} \tag{2}$$

The obtained problem is the interior boundary value Neumann problem. The equality of derivatives under matching conditions makes Neumann problem solvable. To further facilitate the solution, one more assumption is made:

$$z = h \quad \frac{\partial T_1}{\partial r} = \frac{\partial T_2}{\partial r} \Rightarrow q_0 = bh - a. \tag{3}$$

Furthermore, by means of integral transformation for bounded domain [5], the solution can be obtained as follows:

$$\begin{aligned} T_1(r,z) &= \frac{(a-bz)R}{2\lambda}\left(\frac{r^2}{R^2}-\frac{1}{2}\right)+\frac{b}{3\lambda R}z^3-\frac{a}{\lambda R}z^2+const+ \\ &+\frac{R^2}{\lambda}\sum_{n=1}^{\infty}\frac{J_0(\mu_n r/R)}{\mu_n^2 J_0(\mu_n)}\left[\left(\frac{b}{\mu_n}+\frac{kq_0}{\mu_n+kR}\right)\left(e^{-\frac{\mu_n}{R}(h-z)}+e^{-\frac{\mu_n}{R}(h+z)}\right)-\frac{2b}{\mu_n}e^{-\mu_n\frac{z}{R}}\right] \\ T_2(r,z) &= \frac{q_0 e^{-k(z-h)}R}{2\lambda}\left(\frac{1}{2}-\frac{r^2}{R^2}\right)+\frac{2q_0}{\lambda k^2 R}e^{-k(z-h)}+\frac{2bh}{\lambda R}\left(\frac{h^2}{6}-\frac{1}{k^2}\right)+ \\ &+\frac{2a}{\lambda R}\left(\frac{1}{k^2}-\frac{h^2}{2}\right)+const+\frac{R^2}{\lambda}\sum_{n=1}^{\infty}\frac{J_0(\mu_n r/R)}{\mu_n^2 J_0(\mu_n)}\left(\frac{b}{\mu_n}+\frac{kq_0}{\mu_n-kR}\right)e^{-\frac{\mu_n}{R}(z-h)}+ \\ &+\frac{R^2}{\lambda}\sum_{n=1}^{\infty}\frac{J_0(\mu_n r/R)}{\mu_n^2 J_0(\mu_n)}\left[\left(\frac{b}{\mu_n}+\frac{kq_0}{\mu_n+kR}\right)e^{-\frac{\mu_n}{R}(z+h)}-\frac{2b}{\mu_n}e^{-\mu_n\frac{z}{R}}+\frac{2k^2Rq_0e^{-k(z-h)}}{k^2R^2-\mu_n^2}\right] \end{aligned} \tag{4}$$

As is known, interior boundary value Neumann problem is solved within an arbitrary constant. It can be theoretically calculated by establishing some additional conditions, or alternatively can be based on experimental data. Matching condition and additional condition (3) allow to express parameters a and b via heat flux density q_0 and coefficient k.

$$a = \frac{q_0(2+kh)}{kh} \quad b = \frac{2q_0(1+kh)}{kh^2} \tag{5}$$

3. Results and discussion

The most significant information for the problems of thermochemical treatment is the knowledge of temperature gradients, emerging in the heating process across different sections of a lengthy workpiece. Assessment of these gradients and constructing a temperature distribution graph across different sections of a semi-infinite cylindrical anode sample requires setting numerical values of q_0, k parameters and an arbitrary constant. Heat flux density q_0 and coefficient k can be derived from the experimental data of the heat flux density into the end of the protruding part of the anode sample [6]. Conversion from one heat flux density to the other is based on the known problem of a semi-infinite cylinder cooled from the side surface, which is heated with a constant heat flux onto the side of the sample. The value of the arbitrary constant should be selected with regard to convenient comparison of this model with the one previously suggested in [2], let us select the arbitrary constant value so that temperature values in the center of the sample at the boundary between the immersed and protruding parts coincided in both models. Then the numerical values of the parameters will be as follows: q_0 = 48.95 kW/m^2, k = 23.066 m^{-1}, $T(0, h)$ = 757.5 °C. The dimensional and thermophysical characteristics of the sample fit typical steel samples submitted to plasma electrolytic treatment: R = 6 mm, h = 1.5 cm, λ = 40 W/(m·K).

The graph below shows the dependence of the temperature of a heated sample on the vertical and radial coordinates calculated from the model with the set temperature on the anode surface (model 1), and from the model with the set heat flux density on the anode surface (model 2). In the protruding part, the vertical temperature gradient is equal for both models and amounts to 12 °C / mm (figure 2). The first difference in models is observed in the assessment of the vertical temperature gradient in the immersed part of the anode (figure 3). Apparently in model 1, the temperature remains unchanged at first, and only nearly half way of immersion begins to rapidly decrease. Meanwhile average temperature gradient equals 1.9 °C / mm. In model 2 the temperature in the exterior part slowly decreases, gradually gaining the speed of digression. The average temperature gradient reaches 12 °C / mm, i.e. corresponds to the gradient in the protruding part.

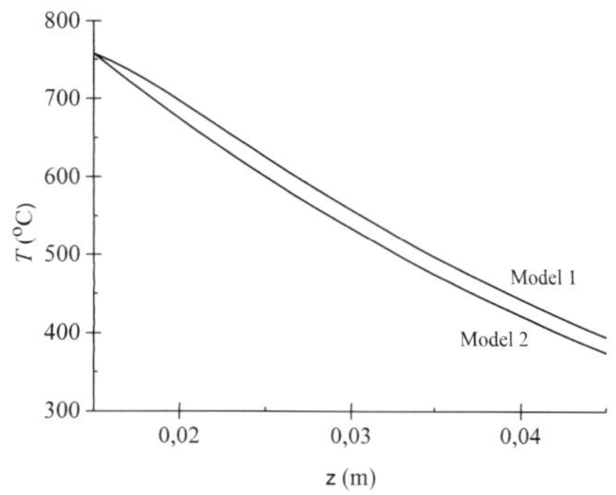

Figure 2. Dependence of temperature on the axis of the sample on the vertical coordinates for the part protruding from the electrolyte surface.

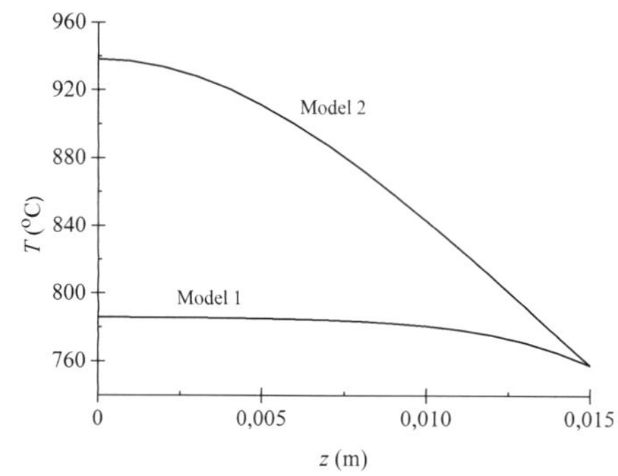

Figure 3. Dependence of the temperature on the axis on the vertical coordinates for the immersed part.

Figures 4 and 5 show temperature dependence on radial coordinate in two cases: on the bottom end of the immersed part, and on the boundary between the immersed and the protruding parts. As mentioned above, in the first model the maximum radial gradient is reached on the boundary between the immersed and the protruding parts and equals 4.5 °C / mm (figure 5), the minimum value of the radial gradient of 0.1 °C / mm is observed on the bottom end of the immersed part. The reverse is true for the second model: the minimum radial gradient is characteristic of the boundary between the immersed and the protruding parts. Whereas the maximum of 3.7 °C / mm is observed on the bottom end of the immersed part (figure 4).

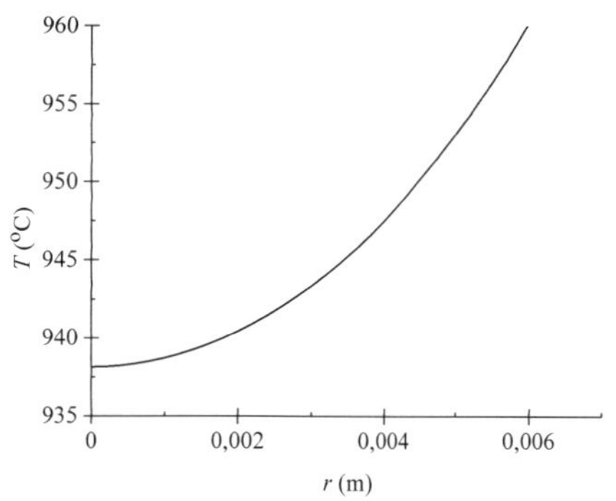

Figure 4. Sample temperature dependence on the radial coordinate on the bottom end of the immersed part for the second model.

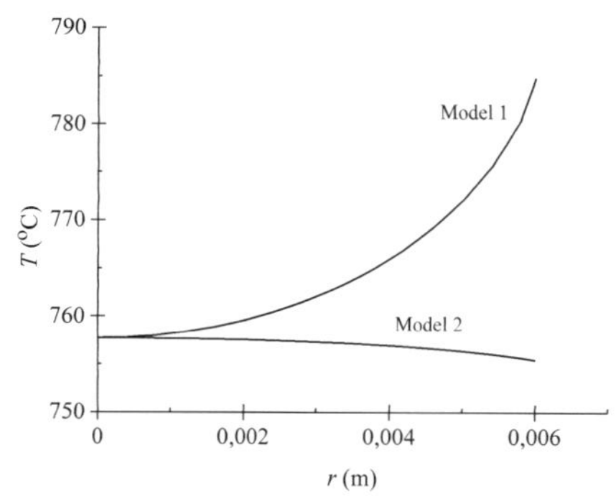

Figure 5. Sample temperature dependence on the radial coordinate on the boundary of the immersed and protruding parts.

The available experimental data [4] allow concluding that the second model is more relevant which can allow determining the sign reversal point in a heat flux on the boundary of the anode and vapouring gas envelope. By means of expression (5) the following is determined:

$$z_0 = \frac{a}{b} = \frac{h(2+kh)}{2(1+kh)}. \tag{6}$$

This expression is not connected with the choice of approximate polynomials as in [4], and is exclusively determined by heat exchange features of the part of a sample protruding above electrolyte surface. When applying the above determined coefficient k heat exchange sign reversal point coordinates equals $0.87h$. Expression (6) implies that at stable heat exchange conditions in a protruding part of the sample, deeper immersion results in a shift of this point to the bottom end of the immersed part.

4. Conclusions

This study offers a model of temperature distribution in a semi-infinite cylindrical sample under anode plasma electrolytic heating in a case of assigned heat flux densities on its surface.

The designed model allows calculating vertical and radial gradients inside a heated anode-sample in its parts immersed in and sticking out from electrolyte. The research shows that the vertical gradient is virtually equal for both parts and is equal to 12 °C / mm. Radial gradient undergoes significant changes along the length of the sample, reaching its maximum of 3.7 °C / mm on the bottom end of the immersed part.

It has been established that the vertical coordinate of the sign reversal point of the heat flux in the immersed part depends on the heat exchange conditions in the protruding part, and is always above half immersion depth of the sample.

Acknowledgments

This work was financially supported by the Russian Science Foundation (Contract No. 18-79-10094) from Kostroma State University.

The authors express their deep gratitude to Prof. S.A. Kusmanov for valuable advice in discussing this work.

References

[1] Belkin P N, Ganchar V I and Tovarkov A K 1986 *J. Eng. Phys. Thermophys.* **51** 154
[2] Belkin P N and Belikhov A B 2002 *J. Eng. Phys. Thermophys.* **75** 1271
[3] Belkin P N, Pasinkovsky E A and Faktorovich A A 1977 *Izv. AN MSSR.* **1** 82
[4] Belkin P N, Mukhacheva T L and Dyakov I G 2008 *J. Eng. Phys. Thermophys.* **81** 1027
[5] Kusse B R and Westwig E 2006 *Mathematical Physics* (Willey-VCH)
[6] Zhirov A V, Belkin P N and Shadrin S Yu 2017 *J. Eng. Phys. Thermophys.* **90** 862

Cathodic nitriding and anodic polishing of Ti6Al4V alloy by plasma electrolysis

I V Tambovskiy[1,2], R A Vdovichenko[2], R D Belov[1], A D Dyakonova[1], R V Nikiforov[1], S A Silkin[1] and I A Kusmanova[1]

[1] Department of Physical, Mathematical and Natural Science, Kostroma State University, 17 Dzerzhinskogo Street, Kostroma 156005, Russia
[2] Department of High-Efficiency Machining Technologies, Moscow State University of Technology "STANKIN", 3a Vadkovsky lane, Moscow 127994, Russia

E-mail: ramstobiliti@gmail.com

Abstract. The paper presents the results of studies on modifying the surface of Ti6Al4V titanium alloy by combined exposure to cathodic nitriding and anodic polishing in electrolysis plasma. The morphology and roughness of the surface, microhardness of the modified layer have been investigated. Wear resistance was studied under dry friction conditions. The effect of combined treatment on corrosion resistance of Ti6Al4V alloy was examined by means of potentiodynamic polarization in Ringer's solution. It has been established that cathodic nitriding at 750 °C for 10 min leads to the hardening of the surface layer up to 820 HV with an increase in roughness by 2 times and wear resistance almost 3 times. Subsequent anodic plasma electrolytic polishing of the nitriding surface in solution of ammonium sulfate leads to a decrease in roughness and friction coefficient with an increase in corrosion resistance.

1. Introduction

Significant attention is paid to improving the performance and aesthetic properties of metal products in various industries. One of the methods of surface modification is the technology of plasma electrolytic chemical-thermal treatment, which is a high-speed method of diffusion saturation of the surface with nitrogen, carbon and/or boron with subsequent quenching.

The plasma electrolytic treatment results of titanium alloys confirm the advantages of this technology. Anodic nitrocarburising of Ti6Al4V titanium alloy leads to the formation of oxide titanium and nitrogen and carbon solid solution in titanium which leads to an increase in microhardness up to 740 HV [1]. The CP-Ti hardness reaches 1500 HV after plasma electrolytic nitrocarburising using the solution containing carbamide, triethanolamine, and formamide [2]. Higher hardness (2000 HV) is observed after nitrocarburising Ti6Al4V alloy in a solution containing of ammonium nitrate, ethanol, and glycerol [3] or 2369 HV using a solution of triethanolamine and formamide [4]. Moreover, the corrosion resistance of CP-Ti in Ringer's solution can increase by means of cathodic plasma electrolytic boriding using borax-based electrolyte due to the formation of nanocrystalline titanium borides [5].

The disadvantage of cathodic processes is a significant increase in surface roughness under the influence of electrical discharges. The roughness parameter Ra increases from 0.179 μm in the control sample to 0.819 in 30 min plasma electrolytic carburising [6] or from 0.4 μm to 5.6 μm after plasma

Content from this work may be used under the terms of the Creative Commons Attribution 3.0 licence. Any further distribution of this work must maintain attribution to the author(s) and the title of the work, journal citation and DOI.
Published under licence by IOP Publishing Ltd

electrolytic nitrocarburising in a solution of ammonium chloride, urea and sodium carbonate [7]. In continuation of the development of technology, additional polishing of the hardened surface in plasma electrolysis – anodic plasma electrolytic polishing (APEP) on the same equipment used for diffusion saturation is proposed. This will allow not only to remove loose areas of the oxide layer, dirt that worsen some of the surface properties, but also to give the product an aesthetic appearance.

2. Materials and methods

Cylindrical samples of Ti6Al4V titanium alloy 11 mm in diameter and 10 mm in height were polished with sandpaper to the surface roughness of $R_a = 1.0 \pm 0.1$ μm and cleaned with acetone in ultrasonic bath. The cathodic plasma electrolytic nitriding (CPEN) was carried out at the sample temperature of 750 °C in an aqueous solution of 5 % ammonium chloride and 5 % ammonia using cylindrical electrolyser under conditions of longitudinal electrolyte flow around the sample. In the upper part of the chamber the electrolyte was being overflowed over the edge into the pallet, from where it was then pumped through the heat exchanger at flow rate of 2.5 L/min. The temperature of electrolyte cooled with tap water was maintained equal to (30 ± 2) °C at the inlet to the electrolyser. During processing, the sample was connected to the negative pole of the DC power supply, and the working chamber to the positive one. After applying a voltage of 220 V, the sample was slowly lowered into the electrolyser at a speed of 1 mm/s to a depth equal to its height. Then the voltage was reduced to 110 V that corresponded to a surface temperature of 750 °C. CPEN was carried out for 10 min, after which the samples were cooled in the electrolyte (quenched). Then samples were washed with water and acetone and dried.

APEP was carried out in aqueous solutions of ammonium sulfate or ammonium chloride of various concentrations at a voltage of 300 V. After the voltage was applied, the sample was slowly lowered into the electrolyzer at a speed of 1 mm /s to a depth equal to its height. The electrolyte was pumped through a heat exchanger at a flow rate of 1 l/min. The electrolyte temperature was maintained at 80±2 °C. The APEP time was 3 min.

The Micromed MET optical metallographic microscope with digital image visualization served to study the surface morphology. The microhardness of the cross sections of the treatment sample was measured using a Vickers microhardness tester (Falcon 503) under a 0.1-N load. 7 indentations were performed on each sample, and 5 points average value except both maximum and minimum values were used for hardness. The surface roughness was measured with the TR-200 profilometer. The change in the weight of samples was determined on the CitizonCY224C electronic analytical balance with an accuracy of ±0.0001 g after washing the samples with distilled water and acetone in an ultrasonic bath to remove traces of salts and subsequently drying.

The wear resistance of samples was studied with the aid of laboratory installation under dry friction conditions with tool alloy steel (wt.%: 0.9–1.2 Cr, 1.2–1.6 W, 0.8–1.1 Mn, 0.9–1.05 C) as counter-body at the normal load of 10 N and sliding speed of 1.555 m/s at the friction path of 1000 m at room temperature. Cylindrical surface of the samples was subjected to wear. Weight loss during the tests was measured with the aid of CitizonCY224C electronic balance after preliminary cleaning of the samples in acetone and subsequently drying.

The potentiodynamic polarisation was carried out to evaluate the corrosion resistance of hardened layers using a Biologic SP-150 potentiostat/galvanostat in Ringer's solution. A conventional three-electrode cell with the bare alloy or treated samples as the working electrode (0.31 cm^2 exposed area), graphite as the auxiliary electrode and a silver-silver chloride reference electrode as the reference electrode, was used. The working electrode was immersed in Ringer's solution for 150 min exposure duration before tests. Then potentiodynamic polarization tests were performed at a scan rate of 1 mV/s. The polarization curves were analyzed to determine corrosion potential and corrosion current density using Tafel's extrapolation.

3. Results and discussion

Cathodic nitriding of Ti6Al4V titanium alloy doubles the microhardness of the alloy (820 HV) and forms a hardened layer with a thickness of more than 300 μm. After processing, the wear resistance of the alloy increases almost 3 times compared to the untreated sample. However, the main disadvantage in the alloy processing technology is the growth of the surface roughness, which after CPEN is 1.9 ± 0.1 μm and is 2 times higher than the initial roughness.

To solve the indicated problem, the possibility of using APEP technology was studied. With APEP in an ammonium chloride solution, an uneven removal of the outer oxide layer from the surface of the nitrided alloy is observed, which leads to the development of a relief and an increase in roughness (figure 1). When processing in solutions containing more than 2 % ammonium chloride, the APEP regime is changed to heating, and the sample is heated to 500–600 °C. At the same time, APEP in solutions of ammonium sulfate makes it possible to reduce the surface roughness of the alloy (figure 2). In this case, the efficiency of APEP depends on the concentration of ammonium sulfate in the electrolyte, with the growth of which it increases. With PEP in ammonium sulfate, the formation of a white oxide film is observed on the surface of the nitrided Ti6Al4V titanium alloy, the development rate of which increases with an increase in the electrolyte concentration, which favorably affects the surface relief.

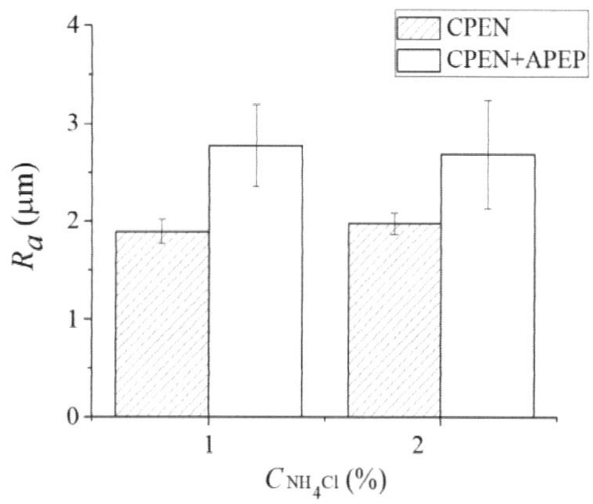

Figure 1. Surface roughness of CPEN samples before and after APEP in solutions of ammonium chloride.

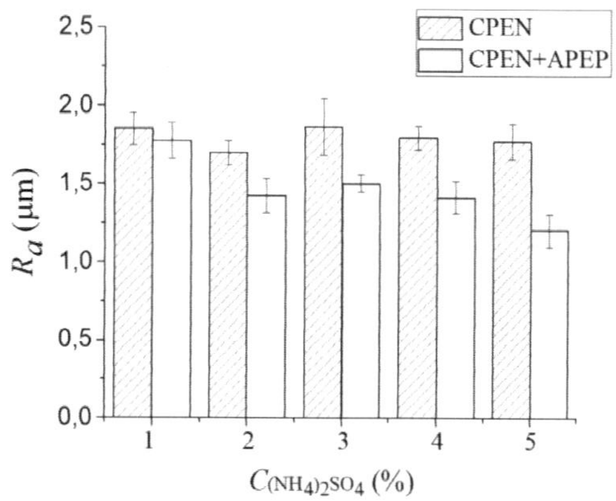

Figure 2. Surface roughness of CPEN samples before and after APEP in solutions of ammonium sulfate.

Tribological tests of Ti6Al4V titanium alloy after nitriding and polishing showed that APEP in 4 and 5 % ammonium sulfate solution reduces the surface friction coefficient (figure 3). However, the weight loss of sample during friction after APEP is always higher than that of a nitrided alloy, and practically does not depend on the concentration of ammonium sulfate in the solution (figure 4).

Table 1 shows the results of potentiodynamic tests of Ti6Al4V titanium alloy before processing and after CPEN and APEP. Complex treatment of the alloy surface leads to an increase in the corrosion current density compared to the untreated way, however, after APEP of nitrided products in ammonium sulfate solutions, a shift in the corrosion potential to the positive region is observed.

4. Conclusions

The possibility of anodic polishing of the Ti6Al4V titanium alloy surface after cathodic nitriding in electrolysis plasma using the same equipment is shown. After CPEN, a modified layer up to 300 μm thick with a microhardness of up to 820 HV is formed with a increase in the surface roughness by 2 times. The wear resistance of the nitriding surface is increased almost 3 times.

APEP the nitriding surface in solutions of ammonium sulfate leads to a decrease in roughness by 1.05–1.5 times and friction coefficient by 1.02 times. In this case, the weight wear becomes more than that of the CPEN sample, but less than that of the untreated one. Polishing of the CPEN surface leads to an increase in the corrosion current density, however, a shift in the corrosion potential to the positive region is observed.

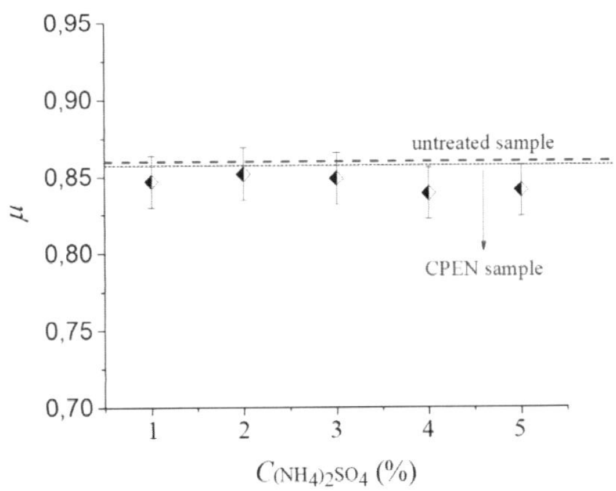

Figure 3. Friction coefficient of CPEN samples before and after APEP in solutions of ammonium sulfate.

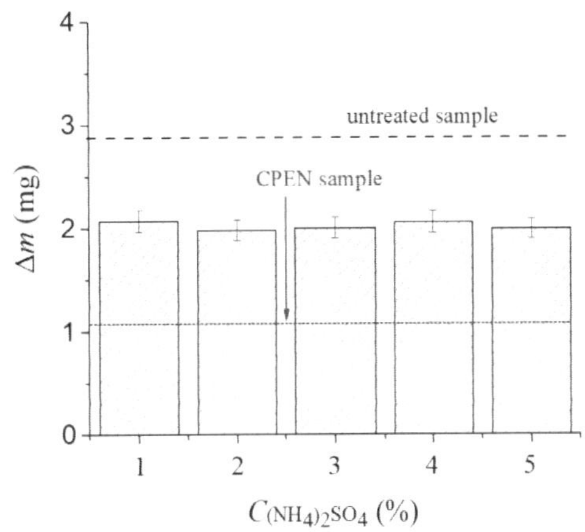

Figure 4. Weight loss of sample at tribological tests of CPEN samples before and after APEP in solutions of ammonium sulfate.

Table 1. Corrosion current density and potential of Ti6Al4V alloy before and after CPEN and APEP in electrolytes with different concentration of NH_4Cl and $(NH_4)_2SO_4$.

NH_4Cl concentration (%)	$(NH_4)_2SO_4$ concentration (%)	Corrosion current density (μA/cm^2)	Corrosion potential (V)
1	–	1.36	-0.254
2	–	3.26	-0.162
–	1	1.40	-0.127
–	2	0.62	-0.084
–	3	0.46	-0.154
–	4	2.12	-0.187
–	5	3.51	-0.118
Untreated sample		0.12	-0.209
CPEN sample		0.58	-0.199

Acknowledgments
This work was financially supported by the Russian Science Foundation (Contract No. 18-79-10094) from Kostroma State University.

The authors express their deep gratitude to Prof. S.A. Kusmanov for valuable advice in discussing this work, and also to Dr. T.L. Mukhacheva for help in analyzing tribological studies.

References

[1] Kusmanov S, Kusmanova I, Tambovskiy I, Belkin P and Parfenyuk V 2019 *Surf. Eng.* **35** 199
[2] Dong Y X, Chen Y S, Chen Q and et al. 2007 *Surf. Coat. Technol.* **201** 8789
[3] Hu Z, Xie F, Liu Y et al. 2008 *Mater. Rev.* **04** http://en.cnki.com.cn/Article_en/CJFDTOTAL-CLDB200804037.htm

[4] Li X M and Han Y 2008 *Appl. Surf. Sc.* **254** 6350
[5] Aliev M Kh and Saboor A 2007 *Bull. Mater. Sc.* **30** 601
[6] Çavuşlu F and Usta M 2011 *Appl. Surf. Sc.* **257** 4014
[7] Kazerooniy N A, Bahrololoom M E, Shariat M H, Mahzoon F and Jozaghi T 2011 *J. Mater. Sci. Technol.* **27** 906

Journal of Physics: Conference Series **2144** (2021) 012034 doi:10.1088/1742-6596/2144/1/012034

Diffusion welding in gradient heatmetry

O A Barabanova[1], S Z Sapozhnikov[2], P A Salmin[1] and S V Nabatchikov[1]

[1] Moscow Aviation Institute (National Research University), 125993, Russia, Moscow, Volokolamskoe sh. 4, Phone: 8 (499) 158-92-09
[2] Higher School of Nuclear and Heat Power Engineering, Saint Petersburg Peter the Great Polytechnic University "SPbPU", 195252, Saint Petersburg, Russia

E-mail: itc-mati@mail.ru

Abstract. Gradient heatmetry allows you to record, process and analyze pulsations of heat flux, which is of paramount importance in the research of gas flow around bodies, in the study of complex heat transfer, etc.

Gradient heatmetry allows you to record, process and analyze pulsations of heat flux, which is of paramount importance in the research of gas flow around bodies, in the study of complex heat transfer, etc.Gradient heat flux sensors (GHFSs) made of materials with anisotropy of thermal and electrical conductivity and Seebeck coefficient have significant advantages. It is especially important that GHFSs have an abnormally high speed: their time constant has a level of 10^{-8} ... 10^{-9} s, which is 3 ... 5 orders of magnitude less than that of the best world analogues. The amount of natural anisotropic metals is small, and it is impossible to influence their properties; in addition, the limitations imposed by their thermal stability are fundamentally unavoidable. For example, GHFSs based on bismuth single crystals of 0.9999 purity have a high and temperature-independent volt-watt sensitivity; however, their field of application is limited by the melting point of bismuth (271^0 C). Therefore, there is obvious interest in artificial anisotropic materials.

To create an artificial anisotropic material, it is required to provide a set of properties that determine the functionality and reliability of the GHFS:

1. Anisotropy of thermal conductivity in two mutually perpendicular directions, attainable with significantly different thermal conductivity of the composite components.
2. Working temperature up to 1300 K and higher.
3. Chemical resistance in various environments.
4. Guaranteed and acceptably strong physical contact in the joint zone of the composite components.
5. Uniform planar and shallow in bedding from the interface front of volumetric interaction, providing a minimum change in the electrical properties of the components of the composite in the joint zone.

Such a composite – a laminated composite material (LCM) – can be obtained by diffusion welding. It allows varying technological parameters (pressure, time, temperature, degree of vacuum) within a wide range, and in the case of a developed and correctly activated surface of the materials to be welded, it guarantees the formation of a reliable connection while maintaining high physical properties of the components to be welded.

Content from this work may be used under the terms of the Creative Commons Attribution 3.0 licence. Any further distribution of this work must maintain attribution to the author(s) and the title of the work, journal citation and DOI.
Published under licence by IOP Publishing Ltd

In the process of diffusion welding, a multilayer anisotropic bar was obtained. After cooling, it was cut into plates located at an angle of 20 ... 45° to the working planes, and then switching leads were attached.

The first composite materials for creating GDTP were LCM nickel + steel 12Cr18Ni9Ti, nickel + steel 65Cr13, chromel + alumel, iron + constantan, and copper + molybdenum. These GHFSs are efficient up to a temperature of 1300°K. The volt-watt sensitivity of some GHFSs at a temperature of about 300 ° K is given in table 1.

Table 1. Volt-watt sensitivity of GHFSs at a temperature of about 300 °K.

LCM	Sensitivity, mV/W
nickel + steel 12Cr18Ni9Ti	0,40
chromel + alumel	0,35
copper + molybdenum	0,02

For further research, we chose LCM nickel + steel 12Cr18Ni9Ti and chromel + alumel. The analysis of the distribution of elements in the joint zone was carried out on a Camebax-Microbeam microanalyzer manufactured by *Cameca* (France). Analysis accuracy is of 5%, spatial resolution: transverse (dia of the analyzed area) is of 1.5–2.5 μm; longitudinal (in depth) is of 0.7 μm.

The microstructure of the LCM nickel + steel 12Cr18Ni9Ti is shown in Figure 1. The results were obtained by electronically scanning the boundary perpendicular to it at a magnification of ×400 (a) and ×2000 (b). The width of the diffusion zone for the studied LCM is of about 1 ÷ 5 μm with a spread of 1 μm, which ensures the required strength of the joint and does not affect the initial materials by 90 ... 95% of their thickness.

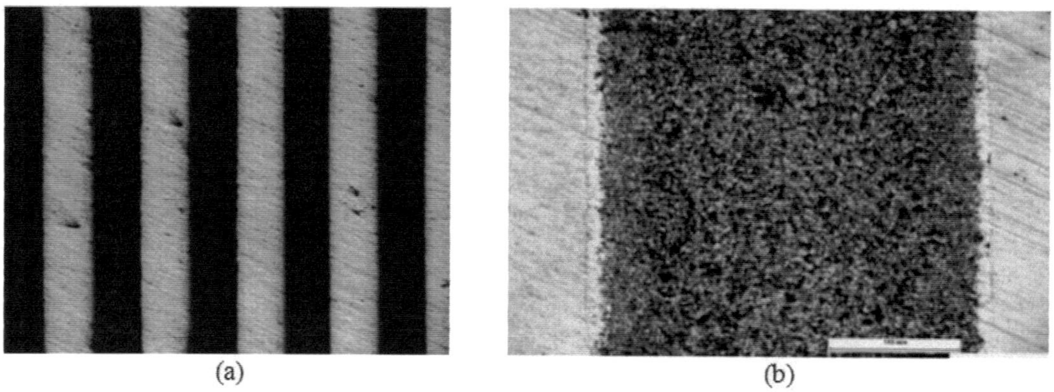

(a) (b)

Figure 1. Microstructure of LCM nickel + steel 12Cr18Ni9Ti with a layer thickness of 0.1 mm: a – before etching (×400), b – after etching of the layers (the diffusion zone is outlined with a dotted line, ×2000).

Figure 2 shows the dependence of the volt-watt sensitivity of the described GHfs on temperature. Composition (a), in which the curves are monotonic, has an advantage.

The calculated volt-watt sensitivity of the GHFS depends on the accuracy with which the properties of the components are specified, but their spread in the literature often exceeds 10 ... 20%. The resulting significant difference between the calculated and experimental curves in figure 2, b indicates the priority of the data obtained during the calibration of the manufactored GHFSs.

Below the results of the calibration of GDTP made from the compositions nickel + steel 12Cr18Ni9Ti and chromel + alumel are shown (table 2).

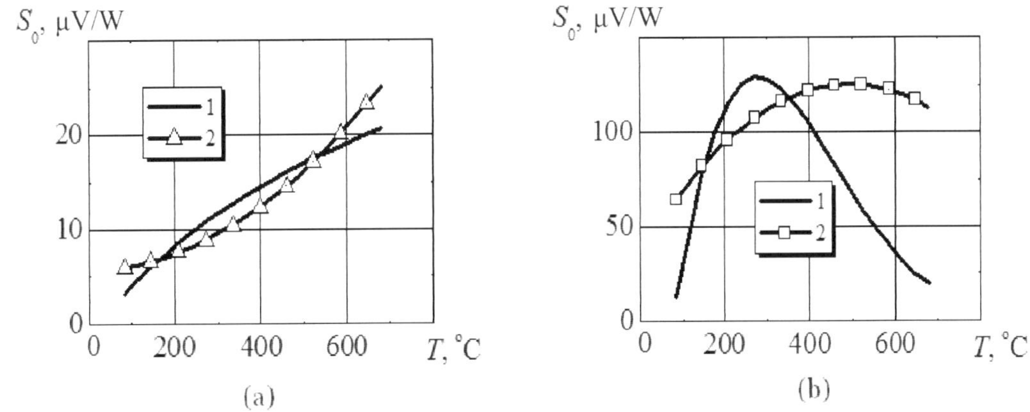

Figure 2. Volt-watt sensitivity of GHFS: a – chromel + alumel; b – nickel + steel 12Cr18Ni9Ti.The numbers indicate: 1 – calculation results; 2 – experimental curves.

Table 2. The results of the GHFSs calibration.

Reference temperature, °C	Heat flux per unit area, W / m^2	Volt-watt sensitivity, mV / W, for compositions		Notes	
		nickel + steel 12Cr18Ni9Ti	chromel + alumel	test time, s	with points processed
104	2600	1,120	0,599	3170	6340
206	5690	1,620	0,562	572	1140
256	9670	1,400	0,461	200	400
336	15200	1,180	0,443	297	594
380	22500	0,844	0,422	281	562

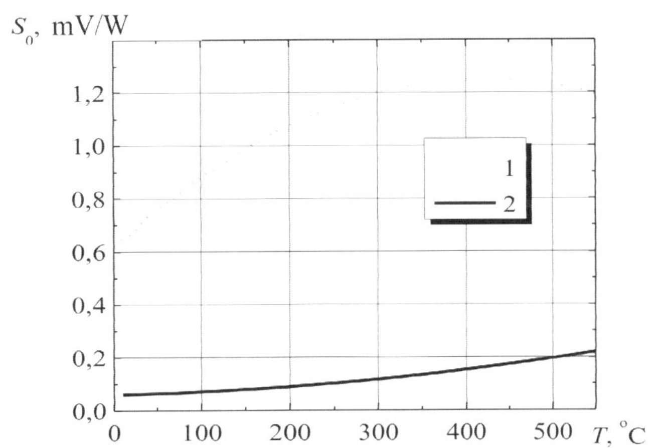

Figure 3. Calibration curves for GHFSs: 1 – nickel + steel 12Cr18Ni9Ti; 2 – chromel + alumel.

From the graph (figure 3) it can be seen that in the temperature range of 250 ... 400 ° C, which is of greatest interest for heat metering in aircraft engines and on the surface of streamlined airfoils, the characteristic of GHFS made of chromel-alumel composition is monotonic, but the sensitivity of sensors based on LCM nickel + steel 12Cr18Ni9Ti is higher almost an order of magnitude, which allows you to prefer this composition.

It was also possible to use permeable materials as layers: fibrous, mesh, with regular perforation, etc. Figure 4 shows the structure of LCM nickel + steel 12Cr18Ni9Ti (mesh).

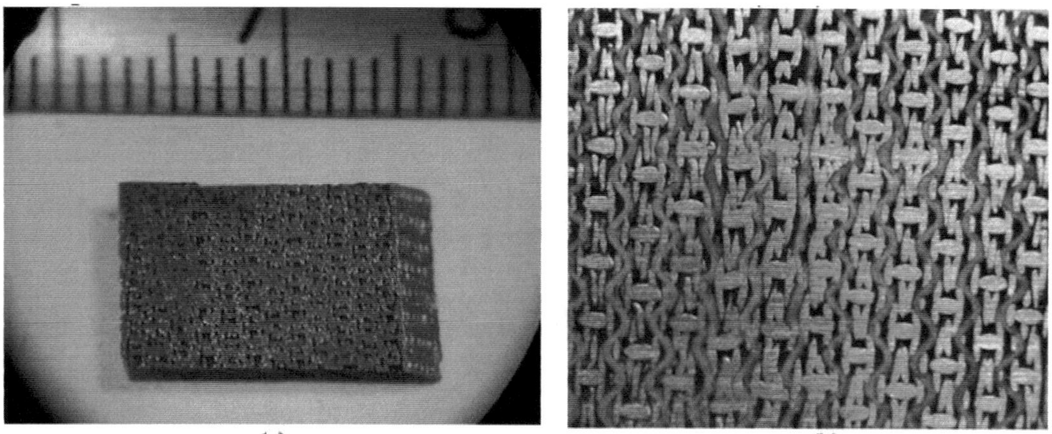

(a) (b)

Figure 4. a) – the sample (scale in mm), b) – microstructure of LCM nickel + 12Cr18Ni9Ti (mesh).

In diffusion welding, the intersections of the mesh threads are welded in the contact zones, and after cutting, the GHFS becomes permeable to liquid or gas flows, and the resistance does not exceed tens of Pa / mm (and can be regulated by the selection of the mesh, the angle of inclination when cutting the bar, and the GHFS thickness). The interpretation of the signal of such a GHFS is a separate problem, but its application in study of heat transfer in systems for blowing a cooled medium, gas curtains, etc. will give clear and unparalleled benefits.

Of particular interest is the use of semiconductor + metal and semiconductor + conductor compositions, since the values of the Seebeck coefficients in semiconductors are an order of magnitude or more higher than the level typical for metals.

Analysis of the literature has shown that sufficient attention has been paid to the preparation of a joint of a single silicon wafer with an aluminum foil; however, the preparation of LCM based on these materials has not been described. By the method of diffusion welding in air (at the temperature of 820 K and holding for 3600 s), it was possible to obtain a GHFS based on silicon + aluminum composition. The high solubility of aluminum in silicon, traditionally used in nonferrous metallurgy, played a positive role, providing a diffusion connection with a transition zone width of 5 ... 15 μm.

It was found that in order to increase the strength of the LCM, welding should be carried out immediately after the activation treatment. An increase in the roughness and (to a lesser extent) microhardness of the surface layer of a single crystal of silicon reduces the strength of the welded joint.

No less promising is a composition of alternating silicon layers with *n*- and *p*-conductivity. We used an aluminum foil of 0.05 mm thickness as interlayers between the silicon wafers and, with the described technology, we obtained a layered *n*- silicon + *p*-silicon composite. The sensitivity of such GHFSs is almost twice the level reached by GHFS silicon + aluminum, and their thermal stability is limited by the softening temperature of silicon (1100 K), however, poor reproducibility of the properties of LCM was revealed, which is largely associated, first, with the method of manufacturing a silicon single crystal (crucible-free zone melting or pulling the crystal out of the melt by the Czochralski method), and, secondly, with the methods of abrasive processing of a single crystal. Both difficulties need to be addressed.

It is important that with the advent of high-temperature GHFSs, both the capabilities and the ideology of gradient heatmetry in aviation and other power plants change significantly.

References

[1] Geiling L 1951 Das Thermoelement als Strahlungdmesser *Zeitung für Angewandte Physik, 3*(12) 467–77
[2] Kazakov N F 1976 Diffusion welding of materials Moscow, Mechanical Engineering 326 p
[3] Diffusion welding of materials: Reference/Edited by N F Kazakov - Mechanical Engineering, 1981 271 p
[4] Kazakov N F et al. 1973 Diffusion connection of silicon with metals. In the collection "Diffusion compound in vacuum" Moscow, 5 pp. 139–46
[5] Krasulin Y L 1971 Interaction of metal with semiconductor in solid phase Moscow, "Science", 119 p
[6] Sapozhnikov S Z, Mityakov V Yu, Mityakov A V, Barabanova O A and Salmin P A 2019 *Russian Aeronautic.* 62(1) 89–95

Possible nature of electrical conductivity of the corund-graphene composite

A G Zholnin and R S Hafizov

The State University of Land Use Planning SULUP, Kazakova str., 105064, Moscow, Russia

E-mail: azholnin@list.ru

Abstract. The effect of sintering modes on the electrical conductivity of the corundum-graphene composite is investigated. The assumption about the nature of the electrically conductive structure and the conditions of its formation is substantiated.

Graphene, which is a two-dimensional alotropic modification of graphite, was obtained by Novoselov and Geim in 2005 [1]. It possesses a number of unique electrophysical, mechanical and thermal properties. Previous studies have shown that the corundum-graphene composite is a promising material both from the point of view of reducing wear during friction and from the point of view of obtaining conductive ceramics with electrical resistance independent of temperature [2–6]. The work [7] shows the promising use of graphene as a hardening additive in ceramic multifaceted cutting inserts for turning tools. In the case of appreciable conductivity of the composite, it becomes possible to electrospark treatment and shape semi-finished products from the graphene / Al2O3 composite. The good electrical conductivity of carbon-ceramic crucibles also makes it possible to use them for heat treatment and electric melting of a number of materials in high-frequency furnaces [8].

The aim of this work is to deepen knowledge about the nature of the formation of a conducting structure when obtaining a corundum-graphene composite, by continuing the research carried out earlier in this direction [9–11].

The corundum-graphene composite was obtained by ultrasonic mixing of graphene powder and aluminum oxide nanopowder in a liquid, followed by their spark-plasma sintering (SPS) or high-temperature pressing (HCP). In the process of compaction, graphene particles are embedded in corundum grains or are located at their boundaries, providing a hardening effect. Under certain compaction modes, it is possible to increase the electrical conductivity of the resulting composite by several orders of magnitude. The electrical resistance was measured at a graphene content of 2 wt. %.

It was found that when sintering a mixture of powders by the HPS method with a pressure of 50 MPa applied during sintering, the presence of graphene had practically no effect on the electrical resistance of the resulting composite. In both cases, a classic insulator was obtained. Its resistance dropped by four orders of magnitude upon heating to the sintering temperature and restored upon subsequent cooling to room temperature.

High electrical conductivity was obtained when sintering was carried out at low pressure (1 MPa). At room temperature, the conductivity of the obtained composite was close to the conductivity of metals (7 orders of magnitude lower than that of corundum without graphene)[9]. At the same time, in the process of reheating and subsequent cooling, the electrical resistance was practically independent

Content from this work may be used under the terms of the Creative Commons Attribution 3.0 licence. Any further distribution of this work must maintain attribution to the author(s) and the title of the work, journal citation and DOI.
Published under licence by IOP Publishing Ltd

of temperature and amounted to 0.25 Ω * m. It is assumed that the electrical conductivity is provided by the formation in the bulk of the composite of a conducting network consisting of graphene cells in electrical contact with each other.

The use of an ECP unit for sintering, where heating is carried out by an external heater, without the use of pulse currents, made it possible to measure the electrical resistance directly during the sintering process (1600 ° C, holding for an hour at this temperature and subsequent cooling). During the heating of the mixture, the electrical resistance dropped, during holding it stabilized, but upon subsequent cooling to room temperature, it dropped by another 3 orders of magnitude and amounted to 0.09 Ω * m, i.e. became 7 orders of magnitude less than that of corundum and composites sintered in ATP under pressure. External examination of the resulting compact revealed the presence of two zones: a dark one in the middle of the sample and a light one in its peripheral area. A point resistance measurement revealed a high conductivity in the central regions and a complete absence of electrical conductivity in the periphery.

It was found that during exposure at a temperature of 1600 ° C, a chemical interaction occurs in the composite between graphene carbon and aluminum oxide with the formation of aluminum carbide and carbon monoxide [10]. The result of this reaction is the formation of a porous coarse-grained structure from corundum grains and aluminum carbide flakes (figure 1). This temperature can be considered as the temperature limit of stability of the corundum-graphene composite.

In figure 2 shows a photograph of the cleavage of the central section of the composite, taken on an SEM without applying a conductive coating, is shown. Electrically conductive areas are darker in color. It can be seen that they form a grid that permeates the bulk of the composite and ensures the electrical conductivity of the entire sample. Dark areas at higher magnification have a columnar structure, which is not typical for corundum (figure 3).

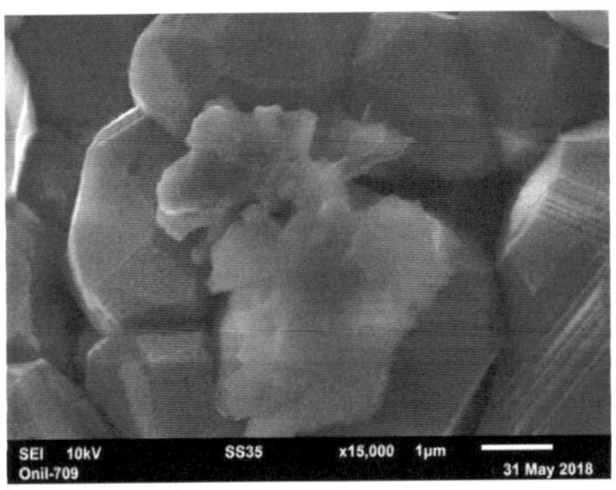

Figure 1. An example of a flocculent formation at the peripheral portion of the sample.

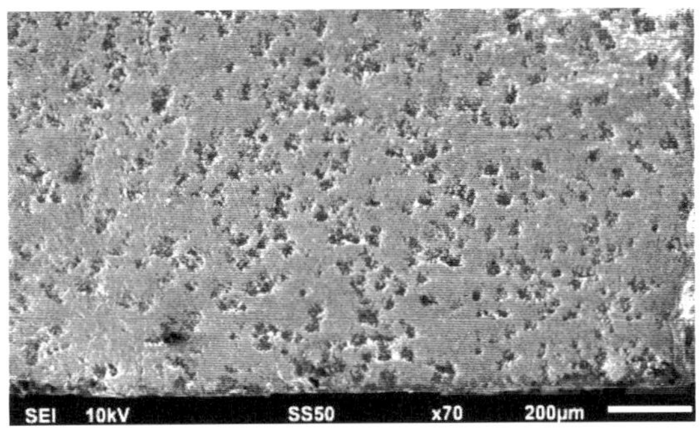

Figure 2. Microstructure of the cleavage of the electrically conductive part of the corundum-graphene composite.

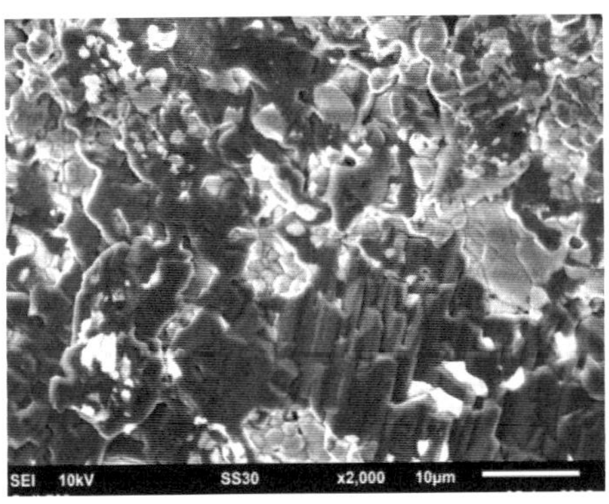

Figure 3. Columnar structure of dark blotches.

The question of the nature of the conductivity of the corundum-graphene composite remains unclear. The electrical conductivity of the composite is caused by the formation of a network consisting of graphene flakes in contact with each other, or graphene forms electrically conductive chemical compounds with corundum, and from them an electrically conductive network is formed, observed in figure 1. The latter is supported by the presence of structures in the conduction zones that are not characteristic of corundum (figure 2), and traces of phases of the complex compound Al_4O_4C detected in X-ray spectra [10].

Thus, previous studies have established:

1. The formation of an electrically conductive structure in the corundum-graphene composite takes place only during sintering at low pressure on the powder.

2. The temperature limit for the existence of an electrically conductive structure in a composite is a temperature of about 1600 ° C, when graphene chemically interacts with aluminum oxide with the formation of aluminum carbide and carbon monoxide.

To clarify the question of at what stage of the formation of a compact the formation of a conducting structure takes place, 5 samples with a diameter of 15 mm and a thickness of 1-1.5 mm were compacted by the SPE method in various modes. In all cases, the graphene content was 2% by weight; heating and subsequent holding were carried out at 1550 ° C. The sintering modes are listed in more detail in the table 1.

Table 1. Sintering modes.

Sample no.	Heating pressure, MPa	Holding pressure, MPa	Holding time, min	Electrical resistance, Ω
1	50	50	10	215
2	0,6	50	10	210
3	0,6	0,6	0,5	220
4	0,6	0,6	10	21
5	0,6	0,6	20	2,8

After sintering, the electrical resistance of the samples was measured at room temperature along their axis. The results are also shown in the table.

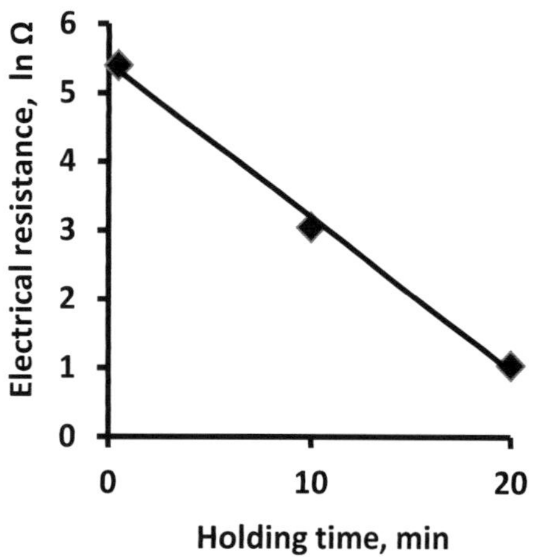

Figure 4. Electrical resistance of specimens sintered at a pressure of 0.6 MPa versus holding time.

From the results obtained it follows:

1. The conductive structure is formed during holding at the maximum temperature only at a pressure of 0.6 MPa in the presence of holding at the maximum temperature. With increasing exposure time, the electrical resistance decreases exponentially (figure 4).
2. Applying a pressure of 50 MPa both at the beginning of heating and during holding blocks the formation of a conductive structure.

From the foregoing, it can be assumed that the conducting structure is formed as a result of a chemical reaction between graphene and aluminum oxide at its initial stage, when compounds of the Al_4O_4C type are formed. For the reaction to proceed, it is necessary to remove carbon monoxide from the reaction zone. At a pressure of 50 MPa, there is no open porosity, carbon monoxide is not removed, and the reaction is blocked. At low sintering pressure, open porosity is present, which ensures the removal of carbon monoxide from the reaction zone. At a temperature of 1600 ° C and an hour exposure, the reaction ends with the formation of aluminum carbide. The conducting structure is destroyed, which we observed in [10].

References

[1] Stankovich S, DiKin D A, Dommet G H et al. 2006 *Nature* **442** 282–6
[2] Reimanis I E 1997 *Mater. Sci. Eng. A* **237** 159–67
[3] Centeno A, Rocha V G, Alonso B et al. 2013 *Journal of the European Ceramic Society* **33** 3201–10
[4] Wan Jiang, Yuchi Fan and Akira Kawasaki 2012 *Adv. Funct. Mater.* 1–7 DOI: 10.1002/adfm.201200632
[5] Yuchi Fan , Lianjun Wang, Jianlin Li et al. 2010 *Carbon* **48** 1743–9
[6] Ting He, Jianlin Li and Lianjun Wang 2009 *Materials Transactions* **50** 749–51
[7] Peretyagin P Yu 2017 *Doct. Diss., Moscow, MGTU Stankin* 184 http://stankin.ru/science/dissertatsionnye-sovety/d-212-142-01
[8] Popov V A 1969 Non-metal materials Handbook **5** Machinebuilding 621.002.33 010727 544
[9] Zholnin A G, Tenishev A V and Stolyarov V V 2019 *Deformatsiya i razrusheniye materialov* **2** 20–7
[10] Zholnin A G, Stolyarov V V, Savel'yev M D and Shornikov D P 2019 *Poroshkovaya metallurgiya:* Inzheneriya *poverkhnosti, novyye poroshkovyye kompozitsionnyye materialy. Svarka. Minsk* 460–9
[11] Bolinches A S, Klyatskina E F, López S A et al. 2018 *Sintez i konsolidatsiya poroshkovykh materialov: (23–26 oktyabrya 2018 g., Chernogolovka)* 21

Development of technological processes for production of items from composite materials based on criterion system

V I Galkin, A R Paltievich and E V Galkin

Department of Technologies and Systems for Computer-Aided Design of Metallurgical Processes in Federal State Budgetary Educational Institution of Higher Education "Moscow Aviation Institute" (National Research University), Volokolamskoe highway 4, Moscow, 125993, Russia

E-mail: tomd@mati.ru

Abstract. The method of development of technological processes of production of items from composite materials is developed, based on application of the criterion system providing formation of strong connection of components and required level of mechanical properties in the material.

1. Introduction

Metal forming under pressure processes (MFUP) are an effective method of producing items and semi-finished products from composite materials (CM) on a metal basis. Using MFUP processes, technologies have been developed for the production of products from fibrous CM (FCM), layered CM (LCM), CM, reinforced with particles. The development of new processes for the production of products from CM is a complex technical task, solving which it is necessary to take into account the peculiarities of the structure and composition of the CM procurement, the possibilities and limitations of production technology, as well as the requirements for the final item.

The manufacturing process of CM products should ensure the formation of a compact structure material with a strong connection of components, maintaining their continuity and initial strength.

2. Task setting

The use of components with different physical and mechanical properties in the CM creates a problem of their compatibility both at the manufacturing stage and during operation.

For reinforced materials with a metal matrix, the problem of physical compatibility is mainly due to the difference in the coefficients of linear thermal expansion at the same pressures and temperatures [1,2]. In the case of LCM, the compatibility problem in the use of materials with a significant difference in physicochemical and mechanical properties.

The interconnection of the main parameters of the process allows to describe a system of criterion expressions linking the characteristics of the workpiece, the finished product and the process parameters into a single system ensuring compliance of the product characteristics with the required indicators.

3. Formalization and problem solving

Depending on the type of CM, the following process criteria can be formulated:

Content from this work may be used under the terms of the Creative Commons Attribution 3.0 licence. Any further distribution of this work must maintain attribution to the author(s) and the title of the work, journal citation and DOI.
Published under licence by IOP Publishing Ltd

1. Compactness criterion.

The CM after the deformation process should have a compact structure - during the compaction of the prefabricated workpiece, its complete compaction should occur. For FCM having a significant initial degree of looseness of the prefabricated blank, such a criterion is the coefficient of looseness K_{np}.

2. Criterion for formation of components strong connection.

The formation of a compound at the interface of the components implies their physicochemical interaction. The strength of the joint is ensured on the one hand by the level of stress-strain state required for these materials, and on the other, by the time of formation of a strong bond [3].

According to the theory of topochemical reactions, the formation of a strong compound for solid-phase processes proceeds in three stages:

$$t_c = t_{ph} + t_a + t_p, \tag{1}$$

where: t_{ph} - is the time of physical contact of the connected components, t_a - is the duration of the activation stage, t_p - is the time of the volumetric chemical interaction.

In the case of FCM, maintaining the necessary level of compaction stress during the time of formation of a strong bond requires the introduction of additional criteria to ensure the safety and continuity of reinforcing fibers [4].

For LCM the compacting stress of prefabricated item, should ensure strong connection of components with formation of metal bond at the interface of layers, but without violation of their initial continuity. To do this, it is necessary to determine the minimum required degree of joint deformation of components ε, which ensures formation of their strong bond σ_{con} [5,6]:

$$\sigma_{con} = f(T, \varepsilon, t_{д}), \tag{2}$$

where: T - is the plastic deformation temperature; ε - extent of deformation; td - deformation time.

Thus, the strength criterion makes it possible to evaluate the speed mode of the deformation process, as well as the optimal degree of deformation of the workpiece.

3. Criteria for maintaining component continuity.

In the process of compacting the FCM blank part, the components experience different levels of force. At the initial stage of the compaction process, the components perceive elastic deformation. Then follows the step of single-component deformation - only the matrix undergoes plastic flow, while compacting the workpiece. At the next stage, all FCM components will be deformed plastically. When designing the technological processes for the manufacture of products from FCM, the appearance of the last stage should not be allowed, since it is accompanied by active crushing of fibers due to their low ductility.

For the matrix component of LCM, the condition of its continuity must be met [6]:

$$\sigma_{i_{con}}^{j} < \sigma_{в_м}, \tag{3}$$

where: $\sigma_{i_{con}}^{j}$ – is the actual (actual) tension arising in component layers i under the influence of external effort at $\varepsilon = \varepsilon_{con}$; $\sigma_{в_м}$ – is the ultimate strength of the matrix component material; j-number of deformation cycle.

According to the second criterion, at the first deformation cycle in the reinforcing component, the condition of its continuity should also be met:

$$\sigma_{i_{con}}^{j} < \sigma_{в_а}, \tag{4}$$

where: $\sigma_{в_а}$ – is ultimate strength of reinforcing component material.

In subsequent cycles, when the ductility margin of the reinforcing component is exhausted, the level of internal stresses in it reaches or exceeds the value of ultimate strength, which indicates the beginning of the process of transferring the layer structure to the particle reinforced structure:

$$\sigma_{i_{con}}^{j} \geq \sigma_{в_а}, \tag{5}$$

Thus, the amount of reduction regulates the transition of the laminate structure to the dispersion-hardened structure.

4. Temperature criterion.

The temperatures at which the CM is formed should correspond to the hot treatment interval of the matrix component pressure, i.e. the following conditions are mandatory [7]:

$$\Delta T_{д} \cap \Delta T_{гд_{м}}, \qquad (6)$$

where: $\Delta T_{д}$; - means temperature intervals of deformation of preparation, respectively; $\Delta T_{гд_{м}}$– is temperature interval of hot deformation of a matrix component.

The above criteria are not exhaustive in the manufacture of CM products. Type of CM, technological process of its manufacture, peculiarities of physical and chemical structure of reinforcing materials, structural features of CM can determine additional parameters in the criterion system, in order to ensure a compact structure with high mechanical and operational characteristics of CM.

4. Experimental testing of solutions

On the basis of the criterion system, there were developed technological processes for the production of pipes from FCM the Al-B system and sheets from LCM of AD1-M1 system.

Boroaluminium pipes were manufactured by drawing on a movable mandrel using a technological shell (figure 1) [8].

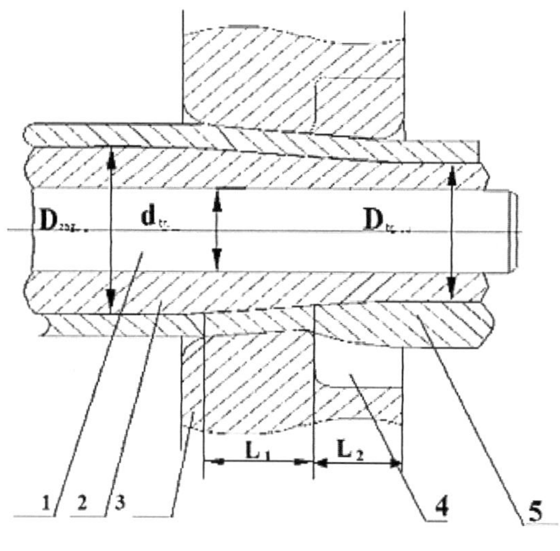

Figure 1. Diagram of production of pipes from FCM by drawing using process shell. 1 - mandrel; 2 - FCM blank part; 3 – die plate; 4 - cavities in the die plate; 5 - process shell.

The process parameters developed on the basis of the critical system ensured a compact structure of the finished product with the required strength characteristics.

For laminated sheets, the criterion approach made it possible to calculate the modes of multicycle rolling of aluminum-copper blank parts (table 1) [8].

The experimental implementation of the LCM rolling method shows that the process criteria enable the transition of the laminated structure of the material to dispersion-hardened on the 4th rolling cycle (figure 2a), and after the sixth cycle (figure 2b), the structure of the obtained material is a matrix in which reinforcing copper particles are evenly distributed, with an average area of these particles of about 714 μm^2.

Table 1. Modes of multicycle rolling of blank parts AD1 - M1

№ of rolling cycle	ε_Σ in a cycle, %	Set deformation (between heatings)	H_o, mm	H_1, mm	n, pieces.	Layer quantity M1, in set, pieces.	Quantity of layers AD1, in set, pieces.
1	55,5	1 pas. - 50% 2 pas. - 5%	4,5	2	11	1	2
2	90,9	1 pas. - 50% 2 pas. - 50% 3 pas. - 37,5%	22	2	8	11	22
3	97,5	1 pas. - 50% 2 pas. - 50% 3 pas. - 50%	16	2	4	88	176
4	75,0	1 pas. - 50% 2 pas. - 50%	8	2	4	352	704
5	75,0	1 pas. - 50% 2 pas. - 50%	8	2	4	1408	2816
6	75,0	1 pas. - 50% 2 pas. - 50%	8	2		5632	11264

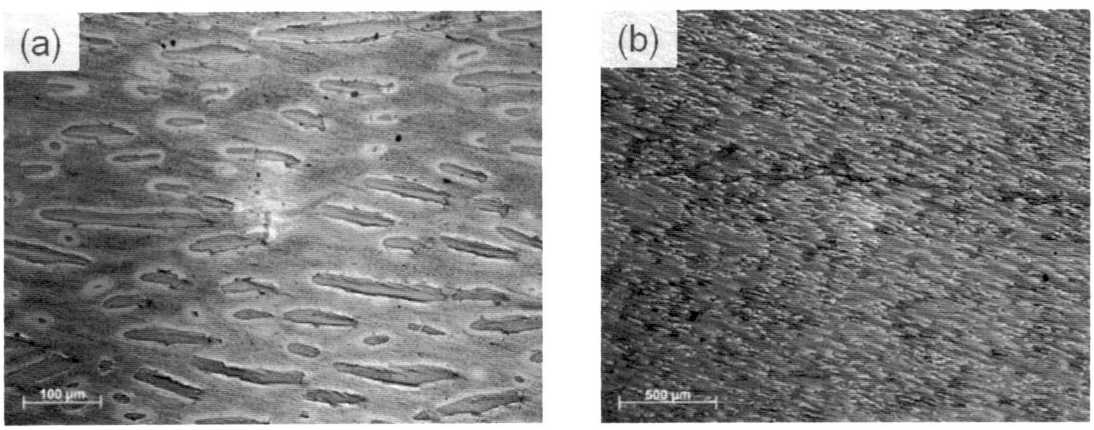

Figure 2. Microstructure of aluminum-copper composite. a - after the fourth cycle of rolling (x200); b - after the sixth cycle of rolling (x50).

5. Conclusion

The proposed methodology, based on the use of a criterion system, can be considered universal and allows to develop technologies of the process of producing CM with various types of strengthening, ensuring the production of products with pre-predicted high mechanical and operational properties.

References

[1] Composites. Under the general ed. of L Brautman and R Krok. 1978 Composite materials with metal matrix. Editor K Kreider, translation from English, under edition of K I Portnoy, Moscow: Engineering 504 p

[2] Auchynnikau Y V, Liopa V A, Voznyakovsky A P, Voznyakovsky A A, Chekan N M and Eysymont E I 2020 *Progressive technologies and systems of mechanical engineering* **2(69)** 63–7
[3] Postnikov V S, Lavrent'ev V I and Shorshorov M Kh 1983 P*hysics and chemistry of material processing* **5** 96–9
[4] Kocheshkov I V 2013 *Production of rolled products* **12** 15–9
[5] Antipov V V, Senatorova O G, Sidelnikov V V and Shestov V V 2012 *Adhesives. Sealants. Technologies* **6** 13–7
[6] Fridlyander I N, Senatorova O G, Lukina N F, Antipov V V, Sidelnikov V V, Grinevich A V and Postnov V I 2005 *Adhesives. Sealants. Technologies* **7** 15–7
[7] Batienkov R V, Burkovskaya N P, Bolshakova A N and Khudnev A A 2020 *Proceedings of VIAM* **6-7(89)** 45–61
[8] Galkin V I, Evseev P S and Galkin EV 2016 *Materials science news. Science and Technology* **3(21)** 4–16

Journal of Physics: Conference Series **2144** (2021) 012037 doi:10.1088/1742-6596/2144/1/012037

Electrolytic superconducting niobium coatings for a cryogyroscope rotor: creation and properties

M A Okunev, A R Dubrovskiy, O V Makarova and S A Kuznetsov

Tananaev Institute of Chemistry of the Federal Research Centre "Kola Science Centre of the Russian Academy of Sciences", 184209 Apatity, Russia

E-mail: m.okunev@ksc.ru

Abstract. Electrodeposition of niobium coatings on spherical substrates for a cryogyroscope rotor made of carbopyroceram was considered. A special design of the installation for obtaining coatings on spherical samples was created. The coatings were applied at a temperature of 750 °C with the cathodic current density of $5 \cdot 10^{-3}$ - $2 \cdot 10^{-2}$ A·cm^{-2} and the electrolysis time of 8-12 h. It was found that the use of a cathodic current density of $2 \cdot 10^{-2}$ A·cm^{-2} and higher is impractical, because there is a roughening of the coatings surface. The composition of electrolytic coatings was identified by XRD analysis. The macrostructure of niobium coatings was studied using electron microscopy. The concentration of metallic impurities in the niobium coatings was defined by spectral quantitative analysis. The content of gas impurities was determined by gas chromatography. The roughness, nonsphericity, and superconductive properties of niobium coatings were investigated.

1. Introduction

In study [1] it was found that the most promising substrate for a cryogyroscope rotor is a carbopyroceram (CPC). The use of a rotor made from this material in a cryogyroscope is possible if a superconducting coating is applied to the substrate. Niobium was used as a superconductor because it has a high critical temperature of the transition to the superconducting state in comparison with other elements. It can also be easily processed by mechanical, chemical and electrochemical methods.

There are many methods for obtaining niobium coatings. Electrolysis of molten salts is the most promising of them, because this method makes it possible to obtain uniform coatings, including articles of complex configuration. Since the rotor is a spherical sample with a diameter of 10 mm, a special design of a rotating cathode was created to accomplish this task. A rotating cathode consists of two molybdenum discs located horizontally and parallel to each other. The upper disc remains stationary and has an annular stop to prevent the sphere from falling, and the lower one is mounted on the vertical rotating shaft of the overhead stirrer. The sample moves between the discs, which makes it possible to obtain coatings uniform in thickness due to electrical contact at any point of the sphere.

For a better understanding of electrodeposition process, the electrochemistry of niobium on a carbopyroceram electrode was studied. The kinetic parameters (transfer coefficient and diffusion coefficient) of the electroreduction process Nb(IV) to Nb in an electrolyte of the composition NaCl-KCl-NaF (10 wt.%)-K_2NbF_7 in contact with a metallic niobium at a temperature 1023 K were obtained.

Content from this work may be used under the terms of the Creative Commons Attribution 3.0 licence. Any further distribution of this work must maintain attribution to the author(s) and the title of the work, journal citation and DOI.
Published under licence by IOP Publishing Ltd

2. Experimental

A glassy carbon crucible SU-2000 brand was used as a container for the melt in the processes of elecrodeposition, a spherical sample of carbopyroceram was used as a cathode, and a crucible lining made of niobium was an anode [2]. The conditions and electrolytic method for obtaining a superconducting niobium coating on the cryogyroscope rotors require the creation of a high-purity atmosphere of an inert gas (argon). The inert atmosphere in the cell was maintained using a pressure regulator and an inert gas reservoir. Electrochemical cell for obtaining niobium coatings on spheres is shown in figure 1.

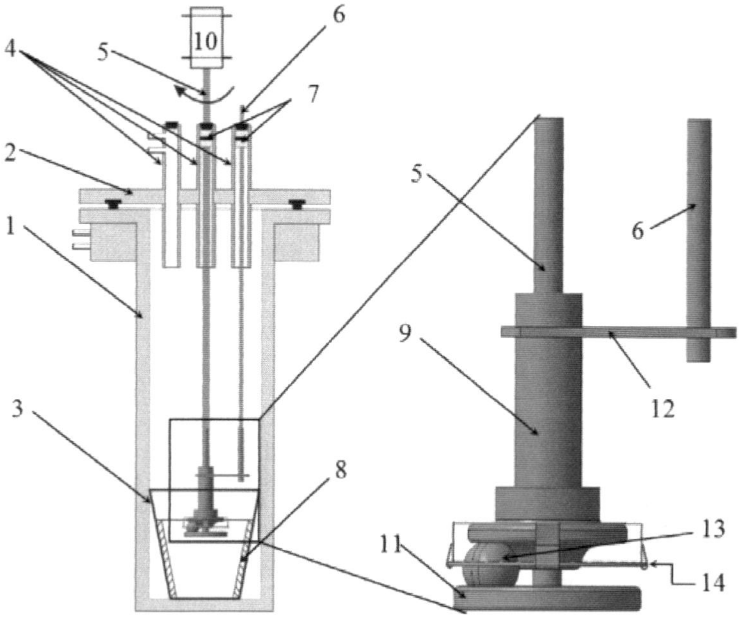

Figure 1. Electrochemical cell for obtaining niobium coatings on spheres. 1 – retort, 2 – lid, 3 – glassy carbon crucible, 4 – holes with spigots, 5 – current lead, 6 – fixing rod, 7 – gasket system, 8 – anode, 9 – cathode, 10 – overhead stirrer, 11 – rotating disc electrode, 12 – retainer, 13 – sample, 14 – ring stop.

The electrodeposition of niobium coatings on spheres was carried out at a temperature of 1023 K. The cathodic current density during electrolysis varied in the range $5\cdot10^{-3}$ - $2\cdot10^{-2}$ $A\cdot cm^{-2}$, the rotation speed of the stirrer in all experiments did not change and was chosen equal to 35 rpm [3]. The electrolysis time was 8-12 h.

Such conditions made it possible to obtain coatings with the thickness up to 140 μm. It was found that the use of a cathodic current density of $2\cdot10^{-2}$ $A\cdot cm^{-2}$ and higher is impractical because there is a surface roughening of the coatings.

For identification of electrolytic coatings a Shimadzu XRD-6000 diffractometer was used. The macrostructure of niobium coatings was investigated using an Axio Observer.D1m electron microscope with a Thixomet image analyzer. The concentration of metallic impurities in the niobium coating was determined by spectral quantitative analysis on an Iskroline 300 instrument and after dissolution of the niobium coating using a Perkin Elmer AAS 4100-ZL atomic absorption spectrometer. The gases in niobium were detected by a Gas Chromatograph-Mass Spectrometer GCMS-QP 2010 Ultra Shimadzu. The roughness and nonsphericity of the samples were measured using a Taylor Hobson Talyrond 595 profilograph-profilometer. The superconducting properties of niobium coatings were determined by an AC-Magnetometer.

3. Results and discussion

The appearance of a carbopyroceram sample before (a) and after electrolysis (b) is shown in figure 2.

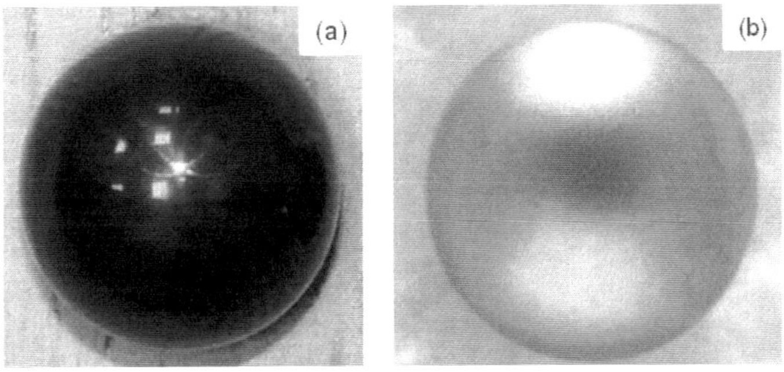

Figure 2. Appearance of a carbopyroceram sample: (a) – initial; (b) – niobium-coated.

The surface of the niobium coating after electrodeposition corresponds to the 8-9 class of surface finishing. Superconducting properties depend on the surface roughness, therefore an electropolishing of a niobium coating is required after electrodeposition. Electropolishing was carried out in a mixture of H_2SO_4:HF (9:1) acids. This treatment allows to reduce the coating roughness.

The concentration of impurities in the niobium coating was analyzed by various methods. Spectral quantitative analysis of niobium coatings was used for detection of the following impurities (wt.%): Mn< $2\cdot10^{-4}$; Mg< $3\cdot10^{-4}$; Si< $1\cdot10^{-3}$; Fe = $2.3\cdot10^{-3}$; Ni< $5\cdot10^{-4}$; Pb< $5\cdot10^{-4}$; Sn< $5\cdot10^{-4}$; Ti< $1\cdot10^{-3}$; Al< $5\cdot10^{-4}$; Co< $1\cdot10^{-3}$; Mo< $1\cdot10^{-3}$; Ca< $1\cdot10^{-3}$; Zr< $2\cdot10^{-3}$; V< $3\cdot10^{-4}$; Cu $\leq 1\cdot10^{-3}$; Cr< $5\cdot10^{-4}$. Gas chromatography was utilized for the determination of oxygen and nitrogen (wt.%): O_2 = 0.01; N_2 <0.001.

The roughness of obtained coatings was R_a = 0.4 μm and nonsphericity was S_p = 0.2 μm.

The superconducting properties of a rotor with electrolytic niobium coating on carbopyroceram were measured in comparison with a rotor made of all-metal niobium (grade "Nb1").

The critical superconducting transition temperature and critical transition fieds for the samples at different temperatures are presented in tables 1 and 2.

Table 1. Critical transition temperature of the sample at "zero field".

Samples	T_c, K	δT_c, K
Nb1	8.84	0.07
CPC	8.36	0.08

Table 2. Critical transition fields of the sample at different temperatures.

Temperature, K	H_{c1}, O_e		H_{c2}, O_e	
	Nb1	CPC	Nb1	CPC
4.5	4.7	7.7	10	~ 12
5.0	4.0	6.2	8.2	9.5
5.5	3.5	5.0	7.0	8.0
6.0	2.8	4.0	6.0	6.2
6.5	2.2	3.0	4.6	4.5
7.0	1.6	2.0	3.5	3.0
8.0	0.6	0.4	1.5	0.8

From the data given in tables 1 and 2, it can be seen that the sample of all-metal niobium is inferior in superconducting characteristics to the sample of carbopyroceram with electrolytic niobium coating. Such characteristics make it possible to use a niobium coating on the carbopyroceram substrate in cryogenic devices.

The final stage in creating of superconducting coating is protection it from the environment because niobium has a high affinity for oxygen. Oxygen reduces the metal plasticity, effects on the electrical properties, increases the specific niobium resistance, and reduces the transition temperature to superconducting state. To eliminate the abovementioned disadvantages, a niobium pentoxide film was formed on the niobium coating. The protective film was obtained by the anodic oxidation in a 1% solution of phosphoric acid. The appearance of a niobium-coated carbopyroceram sample before (a) and after anodic oxidation (b) is shown in figure 3.

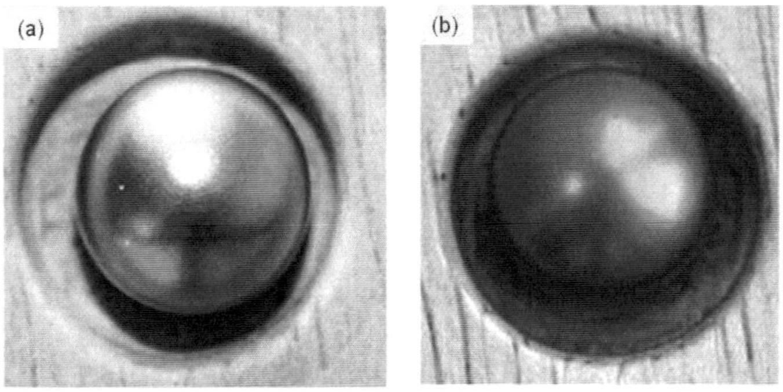

Figure 3. Appearance of a niobium-coated carbopyroceram sample: (a) – initial; (b) – with a protective film.

4. Conclusions

A special design of the installation for obtaining coatings on spherical samples by the electrolytic method was created, which makes it possible to obtain coatings uniform in thickness.

The niobium coatings in carbopyroceram spheres were obtained at 750 ºC at the cathodic current density $5 \cdot 10^{-3}$ - $2 \cdot 10^{-2}$ $A \cdot cm^{-2}$ and the electrolysis time 8-12 h. The rotation speed of the stirrer in all experiments did not change and was chosen equal to 35 rpm. Such conditions made it possible to obtain coatings with the thickness up to 140 µm. It was found that the utilization of cathodic current density of $2 \cdot 10^{-2}$ $A \cdot cm^{-2}$ and higher is impractical because there is a surface roughening of the coatings.

After the electrodeposition the coatings were electropolished in a mixture of H_2SO_4:HF (9:1) acids. This treatment allows to reduce the coating roughness. The treated coating was analyzed for the presence impurities by two methods: spectral quantitative analysis, gas chromatography. The roughness of the coatings and the sample nonsphericity were measured. The superconducting properties of a rotor with electrolytic niobium coating on carbopyroceram were measured in comparison with a rotor made of all-metal niobium.

To solve the problem of niobium protection from the environment a niobium pentoxide film was formed on the niobium coating by the anodic oxidation in a 1% solution of phosphoric acid.

Acknowledgments

The work was carried out with partial financial support of the grant for research projects of young scientists of the Murmansk region in 2021.

References

[1] Dubrovskiy A R, Okunev M A, Makarova O V, Makhaev E A and Kuznetsov S A 2016 *Russ. J. Appl. Chem.* **89** 5 612–8

[2] Dubrovskiy A R, Okunev M A, Makarova O V and Kuznetsov S A 2016 *ECS Transactions* **75** 15 609–16
[3] Dubrovskiy A R, Okunev M A, Makarova O V and Kuznetsov S A 2018 *Coatings* **8** 213

AUTHOR INDEX

Institute of Physics
Dirac House, Temple Back
Bristol BS1 6BE UK

ISSN: 1742-6588
ISBN 978-1-7138-5115-8

9 781713 851158

CURRAN ASSOCIATES INC.
proceedings.com